Wildlife Tourism Futures

THE FUTURE OF TOURISM

Series Editors: **Ian Yeoman**, *Victoria University of Wellington, New Zealand* and **Una McMahon-Beattie**, *Ulster University, Northern Ireland, UK*

Some would say that the only certainties are birth and death; everything else that happens in between is uncertain. Uncertainty stems from risk, a lack of understanding or a lack of familiarity. Whether it is political instability, autonomous transport, hypersonic travel or peak oil, the future of tourism is full of uncertainty but it can be explained or imagined through trend analysis, economic forecasting or scenario planning.

This new book series, The Future of Tourism, sets out to address the challenges and unexplained futures of tourism, events and hospitality. By addressing the big questions of change, examining new theories and frameworks or critical issues pertaining to research or industry, the series will stretch your understanding and generate dialogue about the future. By adopting a multidisciplinary perspective, be it through science fiction or computer-generated equilibrium modelling of tourism economies, the series will explain and structure the future – to help researchers, managers and students understand how futures could occur. The series welcomes proposals on emerging trends and critical issues across the tourism industry and research. All proposals must emphasise the future and be embedded in research.

All books in this series are externally peer-reviewed.

Full details of all the books in this series and of all our other publications can be found on http://www.channelviewpublications.com, or by writing to Channel View Publications, St Nicholas House, 31-34 High Street, Bristol BS1 2AW, UK.

THE FUTURE OF TOURISM: 4

Wildlife Tourism Futures

Encounters with Wild, Captive and Artificial Animals

Edited by
Giovanna Bertella

CHANNEL VIEW PUBLICATIONS
Bristol • Blue Ridge Summit

To Dina, for being my guide into the wild.

To Manuel, for the countless coffee breaks and talks about the futures.

DOI https://doi.org/10.21832/BERTEL8175
Library of Congress Cataloging in Publication Data
A catalog record for this book is available from the Library of Congress.
Names: Bertella, Giovanna - editor.
Title: Wildlife Tourism Futures: Encounters with Wild, Captive and Artificial Animals/Edited by Giovanna Bertella.
Description: Bristol, UK; Blue Ridge Summit, PA: Channel View Publications, 2021. | Series: The Future of Tourism: 4 | Includes bibliographical references and index. | Summary: "This book presents possible future scenarios in wildlife and animal tourism. It offers critically-imagined futures in order to encourage readers to reflect on the possibility of shaping a better future. It will appeal to researchers, students and practitioners in wildlife tourism, environmental studies, sustainability and conservation"—Provided by publisher.
Identifiers: LCCN 2020033486 (print) | LCCN 2020033487 (ebook) | ISBN 9781845418168 (paperback) | ISBN 9781845418175 (hardback) | ISBN 9781845418182 (pdf) | ISBN 9781845418199 (epub) | ISBN 9781845418205 (kindle edition)
Subjects: LCSH: Wildlife watching industry. | Wildlife conservation. | Captive wild animals.
Classification: LCC G156.5.E26 W55 2021 (print) | LCC G156.5.E26 (ebook) | DDC 338.4/759—dc23 LC record available at https://lccn.loc.gov/2020033486
LC ebook record available at https://lccn.loc.gov/2020033487

British Library Cataloguing in Publication Data
A catalogue entry for this book is available from the British Library.

ISBN-13: 978-1-84541-817-5 (hbk)
ISBN-13: 978-1-84541-816-8 (pbk)

Channel View Publications
UK: St Nicholas House, 31-34 High Street, Bristol, BS1 2AW, UK.
USA: NBN, Blue Ridge Summit, PA, USA.

Website: www.channelviewpublications.com
Twitter: Channel_View
Facebook: https://www.facebook.com/channelviewpublications
Blog: www.channelviewpublications.wordpress.com

Copyright © 2021 Giovanna Bertella and the authors of individual chapters.

All rights reserved. No part of this work may be reproduced in any form or by any means without permission in writing from the publisher.

The policy of Multilingual Matters/Channel View Publications is to use papers that are natural, renewable and recyclable products, made from wood grown in sustainable forests. In the manufacturing process of our books, and to further support our policy, preference is given to printers that have FSC and PEFC Chain of Custody certification. The FSC and/or PEFC logos will appear on those books where full certification has been granted to the printer concerned.

Typeset by Deanta Global Publishing Services, Chennai, India.

Contents

	Contributors	vii
1	Introduction: Welcome to the Futures of Wildlife Tourism *Giovanna Bertella*	1
	Part 1: Paths Towards the Futures of Wildlife Tourism	
2	Wildlife Tourism in (Un)sustainable Futures *Qingming Cui*	9
3	Rabbits in the Wild: Close Encounters on an Equal Footing? *Rie Usui and Carolin Funck*	24
4	Representing Wild Animals to Humans: The Ethical Future of Wildlife Tourism *Georgette Leah Burns and Judith Benz-Schwarzburg*	40
	Part 2: Human–Animal Encounters	
5	The Rise of Selfie Safaris and the Future(s) of Wildlife Tourism *Jessica Bell Rizzolo*	57
6	The Future of Captive Wildlife: Useful and Enjoyable for Animals and Visitors? *Ronda J. Green*	71
7	Promises and Pitfalls in the Future of Sustainable Wildlife Interpretation *Gianna Moscardo*	85
8	Interspecies Communication and Encounters with Orcas *Giovanna Bertella*	98

Part 3: Technology Advancements

9 Safeguarding Sustainable Futures for Marine Wildlife
   Tourism through Collaboration and Innovation:
   The Utopia of Whale-Watching 113
   Hindertje Hoarau-Heemstra and Anne-Mette Hjalager

10 Designing Future Wildlife Tourism Experiences: On Agency
   in Human–Sled Dog Encounters 126
   Mikko Äijälä, Titta Jylkäs, Vésaal Rajab and Tytti Vuorikari

11 The Future of Captive Animals and Tourism: The Zoo and
   Aquatic Cloning Centre 2070 140
   Daniel William Mackenzie Wright

12 Will Cryptogovernance Save the Wildlife Tourism Commons? 154
   David Lusseau

13 Final Reflections: Travel Notes, Postcards, Treasures and
   Dragons 167
   Giovanna Bertella

   Index 176

Contributors

Jessica Bell Rizzolo is a postdoctoral research associate at the Department of Fisheries and Wildlife at Michigan State University. She holds a dual PhD in Sociology and Environmental Science and Policy. Jessica's work has appeared in *Society & Animals*, *Crime, Law and Social Change* and numerous edited books. She received a UCLA Animal Law and Policy Grant in 2019 for her work on the wildlife trade. Her research interests include the illegal wildlife trade, wildlife tourism and the intersection of individual animal well-being and species protection.

Judith Benz-Schwarzburg is a senior researcher in ethics and human-animal studies at the Messerli Research Institute of the University of Veterinary Medicine Vienna. She works and publishes at the intersection of animal cognition and animal ethics and has addressed topics as diverse as great ape conservation, animals in entertainment, the representation of animals in picture books, zoo ethics and personhood status in animals. She is currently leading a research group on animal morality. Her book *Cognitive Kin, Moral Strangers?* (Brill, 2019) questions human uniqueness and explores how cognitive kinship matters for animal ethics.

Giovanna Bertella received her PhD from the Department of Sociology, Political Science and Community Planning at UiT – The Arctic University of Norway. She is associate professor at the School of Business and Economics (UiT). Her research interests include: management, marketing, entrepreneurship/innovation, tourism and leisure studies (nature- and animal-based experiences, rural tourism, food tourism), food studies (veganism) and futures studies (scenarios).

Georgette Leah Burns is a senior lecturer in the School of Environment and Science at Griffith University, and a foundation member of both the Environmental Futures Research Institute and Wildlife Tourism Australia. As an environmental anthropologist, her research on ethics in wildlife tourism spans several decades, many continents and multiple species, and has been published in books (e.g. as co-editor of *Engaging with Animals*, Sydney University Press) and journal articles. She is driven by a passion to understand human–animal relationships and enable a future in which people and wildlife equitably co-exist.

Qingming Cui is a researcher in the School of Tourism Management and South China Ecological Civilization Research Center at South China Normal University, Guangzhou, China. He obtained his PhD degree in tourism management from Sun Yat-sen University, China. As a visiting PhD student, he visited the Department of Sociology at The University of Manchester for one year. His research interests include sustainable development, ecotourism, wildlife tourism, food consumption in tourism and tourism social theories.

Carolin Funck obtained her PhD from the Albert-Ludwigs University, Freiburg (Germany). She is professor of human geography at Hiroshima University (Japan), Graduate School of Integrated Arts and Sciences. Her research focuses on the development of tourism in Japan, sustainable island tourism and the rejuvenation of mature tourist destinations; machizukuri and citizen participation are also themes of interest. She is the author of *Tourismus und Peripherie in Japan* and co-author of *Japanese Tourism*.

Ronda J. Green holds a PhD in zoology. Her research includes avian seed dispersal; the effects of habitat alteration on wildlife; and wildlife tourism. She is proprietor of Araucaria Ecotours, specialising in educational wildlife tours, and is currently chair of Wildlife Tourism Australia, for which she has organised and run various workshops, conferences and expos. She has lectured at several universities on ecology and wildlife tourism, co-edited a report for UNWTO on responsible wildlife tourism in Asia and the Pacific, been an invited speaker to conferences in Australia, Brazil, Indonesia and Japan, and travelled in the city and wilderness areas of all non-polar continents.

Anne-Mette Hjalager is a professor at the University of Southern Denmark. Her research interests include innovation in tourism, business development, sustainability, planning and management. She works with policymakers and business associations in Denmark and other countries.

Hindertje (Hin) Hoarau-Heemstra is currently employed as associate professor at the Nord University Business School in Bodø, Norway. Her research interests include management, sustainability, innovation and collaboration. Her main research has been in the context of tourism, focusing on Nordic nature-based tourism and animal-based tourism experiences.

Titta Jylkäs is a service designer and doctoral candidate. She has been carrying out her research at Volkswagen Group and Volkswagen Financial Services in Germany. She is enrolled at the University of Lapland, Finland, in the culture-based service design doctoral programme. In her current research, she focuses on strategic service design in the digital transformation of customer services utilising artificial intelligence.

David Lusseau works on understanding how animals and people reach decisions when uncertain and what the consequences of those decisions are, particularly in intractable situations. He has been in the School of Biological Sciences at the University of Aberdeen since 2007. He obtained is BSc in marine biology at the Florida Institute of Technology and his PhD in zoology at the University of Otago (New Zealand). He then received a Killam fellowship for postdoctoral work at Dalhousie University. He was elected member of the Young Academy of Scotland in 2007, fellow of the Royal Statistical Society in 2009 and fellow of the Royal Society of Biology in 2016.

Gianna Moscardo has qualifications in applied psychology and sociology and joined the School of Business at James Cook University in 2002. Her qualifications in psychology and sociology support her research interests in understanding how communities and organisations plan for and manage tourism development, how tourists learn about and from their travel experiences and how to design more sustainable tourism experiences. Her recent project areas include evaluating tourism as a tool for improving resident quality of life in rural regions, sustainable tourist experience analysis, designing effective tourist interpretation and tourist storytelling.

Véssal Rajab is a service marketing researcher. His passion for creating an ideal customer environment flows through in his latest research contributions with VTT Technical Research Centre of Finland. His research focuses on intelligent transport systems and customer experience in aviation mobility services.

Rie Usui is an assistant professor at the Graduate School of Letters, Hiroshima University. She first conducted wildlife tourism research in 2012 at the Valley of Wild Monkeys in Huangshan, China, where she looked at the role of park rangers in managing tourist–monkey interactions. Since then, her research interests have developed to include a wider range of species in different geographical regions of Japan. More recently, she has been focusing on the animal tourism sites that gained popularity through social media and exploring how the animals and community of people are coping with such a change.

Tytti Vuorikari is a service designer and an independent researcher located in the South-Savo area, Finland. She is a doctoral candidate in culture-based service design at the University of Lapland. In her research, she focuses on real-time documentation of the service design process and self-sustaining innovation spaces.

Daniel William Mackenzie Wright has been studying, researching and lecturing in the field of tourism for 15 years. His career research includes a completed PhD, specifically exploring the role of tourism in post-disaster situations. More recently, the author has focused his research on the tourism futures field; a research area the author believes warrants greater attention. The study of tourism futures has the potential to engage wider audiences. Importantly, highlighting tourism's potential role in the face of many global challenges and opportunities. Dr Wright has presented at international conferences and continues to be an active educator and publisher in tourism-related subject areas.

Mikko Äijälä is a PhD candidate in tourism research at the University of Lapland, Faculty of Social Sciences, Multidimensional Tourism Institute (MTI), Finland. His main research interests include nature-based tourism, human/non-human geographies, human–animal interactions in tourism and multispecies ethnographic methods. Currently, he works at the doctoral school of the University of Lapland and his research focuses on agency in human–sled dog encounters in tourism.

1 Introduction: Welcome to the Futures of Wildlife Tourism

Giovanna Bertella

Terra Incognita: hic sunt dracones
[Unknown Land: here are dragons]
Hunt–Lenox Globe, c. 1500

The core of wildlife tourism concerns humans observing and sometimes interacting with wild animals. The settings where such encounters occur can vary: the tourists can meet the animals in the wild, in protected natural environments and in captive settings (Newsome *et al.*, 2005; Shackley, 1996). The types of encounters can differ: the tourists can observe (e.g. bird watching), touch (e.g. petting zoos), hand-feed (e.g. whale shark feeding) and kill (e.g. trophy tourism) the animals.

For a long time, scholars have investigated the various forms of wildlife tourism in relation to their management as well as their educational and, more recently, ethical dimension (e.g. Carr & Young, 2018; Higginbottom, 2004; Lovelock, 2018; Markwell, 2015). The point of departure for this book is that it might be time to rethink wildlife tourism and adopt a new approach in order to respond to the numerous challenges that the sector faces. Among these challenges, mass wildlife extinction and human population growth can be mentioned as examples (Ceballos *et al.*, 2015). Recently, a certain concern about animal welfare among tourists seems to have emerged, although not always followed by consistent behaviours (Moorhouse *et al.*, 2017; Shani & Pizam, 2008). For wildlife tourism to meet these challenges, new ways of thinking might be necessary.

In the future, wildlife experiences in the wild might be limited. For example, WWF (2020) reports that 17% of the Amazonian forest has been lost in the last 50 years, mostly due to forest conversion for cattle ranching. Without a radical change in food production and consumption, the tourists of the future might only encounter wild animals in captive settings. Furthermore, according to the United Nations' Revision of World Urbanization Prospects (UN, 2018), the proportion of the world's population that lives in urban areas is expected to increase from 55% to

68% by 2050. This trend might influence our perception of what qualifies as 'wild'. Nature can be viewed as a socially constructed concept and so can our definitions of 'the wild' and 'wildlife' (Demeritt, 2002). In an increasingly urbanised society, tourists might view any non-human animal as wild, understood as extraordinary, exotic and fascinating. Signs of this trend are already observable: urban children have very limited experience with farm animals (Knight, 2018). The view of a cow eating grass might be experienced by tomorrow's tourists as wild as today's tourists perceive the view of an elephant.

The necessity to rethink wildlife tourism derives also from technological advancements, for example cloning and virtual reality. The tourism sector might adopt the new technology to develop wildlife tourism giving priority to the tourists' amusement, with little or no concern either for the involved cloned animals or for the implicit educational message of such experiences (Newsome & Hughes, 2017). Another option might be that the animals that today are the main attractions for wildlife tourists would be replaced by virtual animals (Bertella, in press). The application of future technology to the wildlife tourism sector might occur in various ways and with different results but, in any case, would drastically change the sector.

The theme of this book is the exploration of how wildlife tourism might look like in the future. Investigating different types of settings, animals and animal–tourist encounters, this book aims to provide its readers with a broad spectrum of approaches to envision the future, and to provoke deep reflections on relevant aspects such as wildlife tourism sustainability, ethics and management. This can contribute to gaining some new insights on how to proceed in the development and implementation of the type of wildlife tourism that we would like for the coming generations of tourists and, not least, animals.

Methodologically, the idea behind this book is that futures thinking, and in particular scenario developing, might constitute a valuable approach to explore tourism from a new perspective (Bina *et al.*, 2017; Phillimore & Goodson, 2004; Wilson & Hollinshead, 2015; Yeoman, 2019). More precisely, the combination of critical thinking and imagination to envision the future of wildlife tourism is here considered to have great potential. Critical thinking and imagination might allow us to break free from mental barriers, reflect deeply on our relations with wildlife and start building possible pathways towards more desirable futures.

Moreover, the moment we start to think, write and read about critically imagined future scenarios, we also start to consider what it is about the present that we like or fear, and which possibilities we, as scholars, students and general citizens have today in order to shape a better future (Richardson & St. Pierre, 2005). Until now, few scholars have adopted a futures approach to wildlife tourism (Bertella, in press; Wright, 2016,

2018). Thus, this book originates from an invitation to scholars to imagine how wildlife tourism might be in the future. Seventeen scholars answered this call and accepted the challenge to imagine the future of wildlife tourism. Accordingly, this book presents a series of scenarios about the possible futures of wildlife tourism developed by these creative and critical scholars. We, the authors, now invite the readers to follow us while we venture into the future.

In cartography, the term *terra incognita* (Latin: unknown land) was used to indicate unexplored areas that were not yet mapped. We can think about this book as a journey to this still unexplored destination, *terra incognita*. Curiously, an expression that is sometimes reported in association with the term *terra incognita* on ancient maps is *hic sunt dracones* (Latin: here are dragons) suggesting that wild animals might have an important role to play in our future. Dragons are indeed special animals: they are represented as fiery and evil, sometimes benign, non-existent and still ubiquitous and, not least, guardians of treasures (Barnard, 1964; Blust, 2000). The latter feature of dragons about guarding treasures is very promising. At the end of our explorations into the futures described in this book, we might find a treasure. Probably, it will not be gold or jewels, but may come in the form of valuable lessons to apply to the development of wildlife tourism theory and the management of its practice.

Outline of the Book

The future scenarios concerning wildlife tourism presented in this book derive from the application of various perspectives, such as management, innovation, experience design, interpretation and, not least, ethics. These scenarios are grouped into three parts to emphasise their main focus and the book concludes with a final chapter concerning reflections relevant to the future of tourism theory and practice.

Part 1: Paths Towards the Futures of Wildlife Tourism

The depicted scenarios in the first part of the book describe and present some main pathways to the future of wildlife tourism. The key words of these pathways are sustainability, ecofeminism and ecocentrism.

Qingming Cui (Chapter 2) takes inspiration from the debate between green growth and degrowth, science fiction and existing studies about wildlife tourism and conservation. His chapter sheds light on the future of wildlife tourism from a sustainability perspective by picturing different social development paths in response to climate change. Three scenarios are described: green growth, degrowth and dark growth.

Rie Usui and Carolin Funck (Chapter 3) reflect on the role and management of tourism experiences centred on domesticated animals living in the wild. Their chapter is inspired by the case of Ōkunoshima Island (Japan), where the presence of numerous rabbits gone wild attracts tourists and, at the same time, can have negative effects on the rabbits, the local wildlife (flora and fauna) and the safety of tourist–rabbit interactions. This chapter adopts a normative scenario approach and proposes it as a foundation for the management of this and similar challenging situations deriving from the adoption of the main tenets of ecofeminism as a valuable alternative to sustainability.

Part 1 closes with Georgette Leah Burns and Judith Benz-Schwarzburg (Chapter 4) exploring the human–animal relationship in the context of captive wildlife tourism. The focus is on the ethics of this type of human–wildlife encounter. This chapter draws on principles from an ecocentric framework for managing wildlife tourism and ideas from general animal ethics, more specifically zoo ethics. Conservation and welfare, in the context of captive wildlife tourism, are key values for ethically sound discussion of wildlife futures. This chapter proposes a future scenario for captive wildlife tourism to minimise ethical concerns.

Part 2: Human–Animal Encounters

The second part of this book pays particular attention to the encounters between tourists and animals. The desire of tourists to encounter animals is the very core of wildlife tourism and the scenarios in this part of the book explore a variety of possible relations.

Jessica Bell Rizzolo (Chapter 5) considers the recent rise of the selfie safari and the related adverse environmental and animal welfare impacts. Reflecting on the tourism market, and the sociological, cultural and environmental trends that will affect selfie safaris in the future, this chapter constructs a utopian scenario, a dystopian scenario and other plausible scenarios for the futures of selfie safaris.

Ronda J. Green (Chapter 6) examines some of the reasons, good and not-so-good, that wild animals are held captive. Her chapter asks how, in the cases where captivity is probably justified, we can make the animals' lives not only free from suffering but actually pleasurable. Personal observation and reflection from many visits to zoos and wilderness areas across the world, a lifelong habit of animal-watching, both professionally as an ecologist and as a personal hobby, and discussions with zookeepers and others combine with the results of a literature search to imagine what an ideal zoo of the future might look like.

Gianna Moscardo (Chapter 7) uses an intuitive logic and an analytical approach to examine the role of interpretation in the management of wildlife tourist experiences and develops two scenarios. The future of wildlife interpretation is then presented and discussed with particular emphasis on the possibilities provided by information and communication technology and the relevance of engagement and mindfulness in tourism experiences.

Giovanna Bertella (Chapter 8) imagines a future where interspecies communication occurs, and develops three alternative scenarios concerning orca tourism in Northern Norway. The inspirational sources of this chapter include philosophical reflections about animal–human relations, scientific sources about cetaceans, real events and episodes, and fiction. This chapter reflects on how the possibility of mutual communication and understanding might influence wildlife tourism in the future.

Part 3: Technology Advancements

Technology is thought to play an important role in the future of tourism. This aspect has already emerged in some of the chapters in Part 2, and is explored in more detail in Part 3.

Hindertje Hoarau-Heemstra and Anne-Mette Hjalager (Chapter 9) take the development of animal welfare and rights, climate change and marine biodiversity challenges into consideration and develop a utopian scenario for whale watching in the year 2050. Here, technological eco-innovation is discussed in relation to responsible business models and collaboration.

Mikko Äijälä, Titta Jylkäs, Vésaal Rajab and Tytti Vuorikari (Chapter 10) focus on the experiential aspect of wildlife tourism and reflect on the concept of agency applied to animals. Sledding dogs are here viewed as a subject that not only gives the tourists access to the wilderness, but might also give them an insight into pure animal experiences. Underpinned by a speculative design practice, the future scenario presented in this chapter describes how technology might contribute to the recognition of animal agency.

Daniel William Mackenzie Wright (Chapter 11) explores the key themes of zoos and aquariums, animal and habitat extinction, cloning technology and consumer behaviour. He presents a future narrative scenario in a promotional material format concerning a centre where visitors are educated about cloning technology and science, animals and conservation.

Based on the recognition that wildlife has become a key asset in the economy of many countries and the numerous signs suggesting a tragedy of the commons, David Lusseau (Chapter 12) explores how blockchain technology could be used. This adoption is proposed as a possible solution to a key issue in wildlife tourism, which is rights allocation for each time wildlife is used.

Final Reflections

The final chapter of this book highlights its main contributions. Reflecting on the scenarios presented in Parts 1–3, some considerations on possible important lessons about our way to think about and practice wildlife tourism are presented. It is now time to start our journey to *terra incognita*.

References

Barnard, M. (1964) A dragon hunt. *The American Scholar* 33 (3), 422–427.
Bertella, G. (in press) Wildlife tourism in 2150: Uplifted animals, virtual and augmented reality, and everything in-between. In U. McMahon-Beattie, M. Sigala and I. Yeoman (eds) *Science Fiction, Disruption and Tourism*. Bristol: Channel View Publications.
Bina, O., Mateus, S., Pereira, L. and Caffa, A. (2017) The future imagined: Exploring fiction as a means of reflecting on today's grand societal challenges and tomorrow's options. *Futures* 86, 166–184.
Blust, R. (2000) The origin of dragons. *Anthropos* 95 (2), 519–536.
Carr, N. and Young, J. (2018) *Wild Animals and Leisure: Rights and Wellbeing*. London: Routledge.
Ceballos, G., Ehrlich, P.R., Barnosky, A.D., García, A., Pringle, R.B. and Palmer, T.M. (2015) Accelerated modern human-induced species losses: Entering the sixth mass extinction. *Sciences Advances* 1 (5), 1–5.
Demeritt, D. (2002) What is the 'social construction of nature'? A typology and sympathetic critique. *Progress in Human Geography* 26 (6), 767–790.
Higginbottom, K. (2004) *Wildlife Tourism: Impacts Management and Planning*. Altona Melbourne: Common Ground and Sustainable Tourism CRC.
Knight, R. (2018) Quarter of inner city children have never seen a cow, a tractor or come face to face with a hen. *Mirror*, 22 June. See https://www.mirror.co.uk/news/uk-news/quarter-inner-city-children-never-12767926 (accessed 15 August 2020).
Lovelock, B. (2018) *Tourism and the Consumption of Wildlife: Hunting, Shooting and Sport Fishing*. London: Routledge.
Markwell, K. (2015) *Animals and Tourism: Understanding Diverse Relationships*. Bristol: Channel View Publications.
Moorhouse, T., D'Cruze, N.C. and Macdonald, D.W. (2017) Unethical use of wildlife in tourism: What's the problem, who is responsible, and what can be done? *Journal of Sustainable Tourism* 25 (4), 505–516.
Newsome, D. and Hughes, M. (2017) Jurassic World as a contemporary wildlife tourism theme park allegory. *Current Issues in Tourism* 20 (13), 1311–1319.
Newsome, D., Dowling, R.K. and Moore, S.A. (2005) *Wildlife Tourism*. Clevedon: Channel View Publications.
Phillimore, J. and Goodson, L. (2004) Progress in qualitative research in tourism. In J. Phillimore and L. Goodson (eds) *Qualitative Research in Tourism: Ontologies, Epistemologies, Methodologies* (pp. 21–23). London: Routledge.
Richardson, L. and St. Pierre, E.A. (2005) Writing: A method of inquiry. In N.K. Denzin and Y.S. Lincoln (eds) *The Sage Handbook of Qualitative Inquiry* (pp. 959–978). Thousand Oaks, CA: Sage Publications.
Shackley, M. (1996) *Wildlife Tourism*. Boston, MA: International Thomson Business Press.
Shani, A. and Pizam, A. (2008) Towards an ethical framework for animal-based attractions. *International Journal of Contemporary Hospitality Management* 20 (6), 679–693.
UN (2018) Revision of World Urbanization Prospects. See https://www.un.org/development/desa/en/news/population/2018-revision-of-world-urbanization-prospects.html (accessed 15 August 2020).
Wilson, E. and Hollinshead, K. (2015) Qualitative tourism research: Opportunities in the emergent soft sciences. *Annals of Tourism Research* 54, 30–47.
Wright, D.W.M. (2016) Hunting humans: A future for tourism in 2200. *Futures* 78, 34–46.
Wright, D.W.M. (2018) Cloning animals for tourism in the year 2010. *Futures* 95, 58–75.
WWF (2020) Deforestation and forest degradation. See https://www.worldwildlife.org/threats/deforestation-and-forest-degradation (accessed 15 August 2020).
Yeoman, I. (2019) Does the future have a recipe? *Journal of Tourism Futures* 5 (1), 3–4.

Part 1
Paths Towards the Futures of Wildlife Tourism

2 Wildlife Tourism in (Un)sustainable Futures

Qingming Cui

Humans inhabit an increasingly uncertain and unstable world (Bostrom & Cirkovic, 2008), facing existential risks from natural disasters, such as supervolcanic eruptions, and human-caused disasters, including nuclear war and climate change (Currie & Ó hÉigeartaigh, 2018). Bostrom and Cirkovic (2008: 1) pointed out that taking a broad and global view on these existential risks 'allows us to gain perspective and can thereby help us to set wiser priorities'. Future studies provide tools and knowledge to help today's social actors prepare for disasters (Bell, 2003).

It is also insightful to examine tourism futures on a global scale and from an existential risk perspective. The tourism industry is sensitive to the political and economic environment, and tourism development cannot be considered in isolation from macro social situations. Whether the future global community achieves sustainability will directly influence the forms of future tourism. This chapter imagines the futures of one type of tourism – wildlife tourism – from a social sustainability perspective by picturing different social development paths in response to climate change, which is possibly the most urgent and threatening existential challenge facing contemporary societies.

Existing research on wildlife tourism has focused mainly on tourist–wildlife interactions in the present and at a meso-micro scale rather than on a future perspective and a global scale. In contemporary society, tourists encountering wildlife expect a high-quality experience (Dybsand, 2020; Pratt & Suntikul, 2015). Tourist scenic areas and travel agencies often use the opportunity to get close to wildlife as a marketing strategy (Hill *et al.*, 2014). Gazing at, feeding, getting proximity to, touching, even hunting constitute some of the most common desires of modern tourists in encountering wildlife. In response to these demands, the wildlife tourism industry provides various activities that can be divided into three different types: consumptive (e.g. fishing and trophy hunting), low consumptive (e.g. visiting zoos and aquariums) and non-consumptive (e.g. birdwatching and whale watching) (Castley, 2016; Duffus & Dearden, 1990).

The forms of wildlife tourism are embedded in social and historical contexts. What modern tourists desire in their wildlife encounters is influenced by the broader picture of human–wildlife relationships. From this broad perspective, humanity has gradually separated from wildlife. Wild animals have been defeated by humans in the competition for habitat and have therefore been forced to retreat into protected areas. This spatial separation has been accelerated in the process of modernity (Steffen *et al.*, 2015), which has made wildlife, and especially the more endangered species, a novel attraction for urban residents. Hence, the encounter value of wildlife is more appreciated by the tourism industry than its use value and exchange value (Barua, 2017).

Climate change places future human–wildlife relationships in question and creates uncertainties for the wildlife tourism industry. It is said that 'a sixth mass extinction is already under way' (Ceballos *et al.*, 2015: 1). Climate change is expected to accelerate biodiversity loss (Urban, 2015). Moreover, the international tourism industry is itself a significant contributor to global heating (Lenzen *et al.*, 2018). Wildlife tourism, especially the non-consumptive type, usually involves long-distance travel to peripheral areas. Future social responses to climate change will necessarily affect people's attitudes to wildlife and tourism, and thus influence the future forms of wildlife tourism. Predicting how climate change will affect future wildlife tourism is therefore crucial to the future of the industry.

Climate Change and Social Futures

The Anthropocene and climate change

From 1880 to 2012, the average global surface temperature increased by 0.85°C (IPCC, 2013), and the temperature is predicted to keep rising over the next decades based on an analysis of the current global situation. The predicted temperature increase will intensify social and ecological risks and may have catastrophic consequences (IPCC, 2014).

In recent climate change discourse, three basic viewpoints have been expressed (Urry, 2015). *Gradualism* takes the position that climate change is a slowly unfolding result of anthropogenic causes and global development should be adjusted to control the effects. *Scepticism* questions the anthropogenicity of climate change and its catastrophic consequences, with some sceptics holding that climate change is a natural process. *Catastrophism* argues that gradualists underestimate the impacts of climate change and overlook the possible irreversibility of climate forces (Urry, 2015). Among the tourism research community there is some scepticism (e.g. Shani & Arad, 2014), but most scholars concur that global warming is a human-generated result (Hall *et al.*, 2015).

Earth is regarded as having entered a new geological epoch – the Anthropocene – characterised by humanity as a force shaping nature (Crutzen, 2002). Steffen *et al.* (2015) suggest that the Anthropocene started around the mid-20th century, as revealed in graphs showing the 'Great Acceleration': the correlative growth between social economic trends and Earth system trends. During the Great Acceleration, the main contributor to global warming is greenhouse gas (GHG) emissions from social production, movement and consumption, which are powered by fossil fuels. Urry (2013) summarises the 20th century as the high-carbon century, fuelled mainly by oil. The oil civilisation has meant that the entire social system has co-evolved with oil use (Urry, 2013). To reduce the global heating risks, the world needs to systematically transform from a high-carbon society to a low-carbon society (Urry, 2015).

Recently, scholars have argued for two competing pathways to a low-carbon society: green growth and degrowth (Sandberg et al., 2019). The green growth model aims to decouple economic growth from environmental degradation (Fletcher & Rammelt, 2016; UNEP, 2011). Technological and market innovations are regarded as crucial tools to improve the efficiency of the use of natural resources (UNEP, 2011). However, many scholars have criticised the feasibility of green growth, noting its lack of empirical support (Wiedmann et al., 2015), and have proposed an alternative path of 'degrowth' (Kallis, 2011). Degrowth suggests 'repoliticising' society through a fundamental social transformation (Büchs & Koch, 2019). Proponents of degrowth argue that social production and consumption should be downscaled to mitigate GHG emissions (Kallis, 2011). The degrowth proposal is criticised for lacking policy support. Despite the divergence, both of these strategies ultimately accept a gradualist position, accepting that there is a chance to change the situation. The debate over these pathways is ongoing and far from achieving a consensus (Sandberg et al., 2019).

John Urry's scenarios of social futures

John Urry (2013, 2016) proposes some striking scenarios for possible social futures. For him, four different social futures could arise in response to the problems of carbon emissions and climate change. The first scenario is based on the invention of new technologies that solve the problems of energy scarcity and GHG emissions. The social world transforms and facilitates new socio-technical systems to support these innovations. In this future, the world can still embrace prosperity and people are still able to travel using multiple means of transport.

The second scenario predicts a digital world in which digital communications and experiences replace face-to-face networks and the physical production-transport-consumption system. Three-dimensional printing becomes a widespread technology, allowing people to produce their goods at home or in their local neighbourhoods. This digitalised future would achieve huge savings on the costs and energy use associated with transportation.

The third scenario is a 'new medieval' world. This assumes that social actors do not launch any fundamental social transformations to prevent climate catastrophe. Different countries compete for scarce energy resources and do not shy away from using violence to solve interest conflicts. In this future, rich nations and cities are fortressed from the periphery, which Urry (2016) calls the 'wild zone'. Long-distance travel becomes dangerous and infeasible because of its high energy demands; only the super-rich have the capital and resources to visit distant places. People who live in the wild zone face many existential risks, such as food and water shortages, harmful living environments and a violent social atmosphere.

The fourth scenario foresees a low-carbon society achieved through reducing energy use and downscaling production and consumption. Even though overall income may be lower, basic human well-being is ensured. Generally, people live localised rather than mobile lives.

The first and second scenarios depict the adoption of a green growth development path, the fourth scenario is the outcome of a degrowth path and the third path represents a failure to address recent climate and energy challenges, thus allowing conditions to worsen. With the many challenges involved in fulfilling the green growth and degrowth projects for a sustainable future, Urry (2016) believes that the third scenario is more likely to represent the true future for humanity.

Methodology

This study uses scenario planning as a method to envision wildlife tourism futures. A scenario is defined as 'a set of hypothetical events set in the future constructed to clarify a possible chain of causal events as well as their decision points' (Amer *et al.*, 2013: 23). There are four scenario archetypes for the future: continued growth, collapse, steady state and transformation (Amer *et al.*, 2013). Taking inspiration from these archetypes, Urry's scenarios and the debate between green growth and degrowth, this study proposes three potential social future scenarios: green growth, degrowth and dark growth. Green growth corresponds to the transformation archetype, aiming to sustain society through technological change. Degrowth seeks a steady state by downscaling social production and consumption to balance the economy and nature. Dark growth chooses to maintain or increase current growth patterns and follows the inexorable pathway to the collapse of civilisation.

Although Urry makes provocative elaborations of possible social futures at a holistic level and predicts travel and mobility patterns, his scenarios tell us little about tourism and nothing about wildlife tourism. In addition to considering the holistic social conditions sketched by Urry, I plan the future of wildlife tourism by considering three aspects: the state of wildlife conservation, means of travel and human desires regarding tourism and wildlife. First, wild animals are the attractions of wildlife tourism, so the survival and abundance of wildlife determines the range of animal species that people can encounter. According to a recent report, human activities will lead to a maximum of 1 million plant and animal species facing extinction over the coming decades (Tollefson, 2019). However, this trend is not a destiny: situations can change and different future scenarios present different wildlife survival conditions. Second, from 2009 to 2013 the international tourism industry contributed 8% of GHG emissions (Lenzen et al., 2018). In coping with and ameliorating the effects of climate change, the transformation of the means of high-carbon travel is essential. High- and low-carbon societies offer different options for mobility and different levels of accessibility to outlying wildlife destinations. Third, social transformation may involve the transformation of people's attitudes and behaviour in relation to means of travel and biodiversity conservation. Modern tourism stimulates urban people's curiosity to seek novelty in other places (Mitas & Bastiaansen, 2018). In the future, in accordance with different social and environmental ideologies, people may either enlarge or restrain their curiosity towards faraway places.

I take inspiration from multiple sources in proposing forms of future wildlife tourism. First, Urry's future scenarios and the debates surrounding green growth and degrowth (Büchs & Koch, 2019; Fletcher & Rammelt, 2016; Sandberg et al., 2019) help to sketch the macro development paths and social conditions. Second, fiction is an important source of inspiration in presenting predictions and warnings about the future (Bina et al., 2017). Science fiction such as *Interstellar*, *Snowpiercer*, *Elysium* provides details about possible social structures and environmental ideologies that supplement the theoretical imagination on every sketched path. Third, the existing studies on wildlife tourism, including the classification of wildlife tourism (Alves & Lechner, 2018; Castley, 2016), wildlife conservation modes (Di Minin et al., 2016; Duffy et al., 2019) and tourism structures, are the basis from which to project the future. In particular, the consumptive/non-consumptive typology shapes my elaboration on future wildlife tourism. Fourth, scientific reports about biodiversity loss (Maxwell et al., 2016; Urban, 2015) and climate change (e.g. Carleton & Hsiang, 2016; IPCC, 2014) provide the possible conditions for future wildlife, environment and society, which are external changes causing the internal evolution of wildlife tourism. For each type of macro social development path, the knowledge of wildlife, wildlife tourism and tourism in general can be supplemented with potential external changes to deduce the possible forms of future wildlife tourism.

Three Scenarios of Future Wildlife Tourism

Scenario 1: Green growth to a sustainable future

The first scenario predicts a green growth future. Under this scenario, humanity achieves a balance between economic growth and environmental cost through innovation and transformation. According to the mainstream ideas of the United Nations Environment Programme (UNEP, 2011), this balance can be attained by resource decoupling and impact decoupling. Resource decoupling means 'reducing the rate of use of (primary) resource per unit of economic growth' (UNEP, 2011: 4) and impact decoupling means 'increasing economic output while reduce negative environment degradation' (UNEP, 2011: 4). In other words, the green growth path aims to use fewer resources to sustain a larger population with little environment degradation.

In this social vision, humans can still enjoy the benefits of prosperity. The UN's Sustainable Development Goals are largely realised. Only a small percentage of the population continues to experience poverty and hunger, with most people in the world living a healthy, stable and secure life with longer life expectancies. New clean energy is invented, inequality between genders and between countries is reduced and more developing countries attain a developed status.

Green growth is a social and economic change mainly fuelled by material innovation, including technological and market innovation, rather than by a change of values. Thus, with no fundamental change in the environmental ethics of most people.

Because of high productivity and greater efficiency in using natural resources, more wildlife habitats can be restored and conserved as protected areas. The trend of the 'sixth extinction' is reversed and biodiversity starts to revive from its massive losses. However, the expansion of protected areas means that larger financial budgets are demanded for management, scientific monitoring and research. Tourism development is one way to raise funds for conservation. More people live in urban areas due to continued urbanisation processes. As the economy keeps growing, the upper-middle classes can still afford to satisfy their curiosity and escape to wilderness areas. Therefore, in a green growth future, there is space in the market for wildlife tourism products to satisfy the human desire for encounters with nature and wild animals. Most wildlife tourism in protected areas is non-consumptive. The invention of clean energy means that fast, long-distance travel can be sustained with lower GHG emissions. Tourists can use comfortable and rapid means of transport to arrive at remote protected areas.

In terms of consumptive wildlife tourism, trophy hunting becomes a high-end and more ethical niche product. In today's society, trophy hunting is a controversial topic. Some scholars defend trophy hunting because it can raise more money for conservation while contributing less GHG emissions than regular ecotourism (Di Minin *et al.*, 2016). In the green growth scenario, under financial pressure, countries in remote areas with low accessibility have a motive to provide consumptive wildlife tourism activities for rich tourists. However, the regulation of hunting is strict in accordance with ethical considerations, such as a concern for animal welfare and the avoidance of killing animals through maltreatment.

Low-consumptive wildlife tourism, such as visiting zoos, aquariums and other wildlife-based theme parks, still takes place in urban and suburban areas. Zoos and aquariums have important functions in education, the preservation of species, research and recreation (Alves & Lechner, 2018). Although modern zoos and aquariums draw much criticism from proponents of animal rights and animal welfare (Bostock, 2003), they have advantages in providing educational opportunities for urban residents. Hence, zoos and aquariums have not disappeared in the green growth scenario but have improved their animal welfare and environmental interpretation and remain an important recreation space for urban populations. Moreover, because green growth is based on economic innovation, the prevailing social ideology of the market as the solution to environmental problems means that societies have not banned transnational wildlife trade and tourists can watch exotic wild animals in local zoos and aquariums.

Not everyone in the green growth future has an equal opportunity to access all types of wildlife tourism. There is social stratification among wildlife tourists. Generally, future wildlife tourists come from rich urban and suburban areas in developed countries. People who live in rural areas do not often travel and show less interest in wildlife and nature. Urban and suburban residents from the upper classes can afford long-distance travel to outlying areas. Furthermore, they are well educated and more likely to be moral tourists preferring non-consumptive wildlife tourism activities. Middle-class residents often travel short to medium distances to wildlife tourism destinations, including urban zoos, aquariums, theme parks and some protected areas. The urban poor have few opportunities to visit remote protected areas, and they generally visit urban zoos and aquariums to satisfy their desires to encounter exotic wild animals.

However, digital innovations may change the stratification in wildlife tourism consumption. New digital technologies, such as artificial intelligence, virtual reality and holograms, can vividly represent the real natural environment and animals, either at home or in nearby experience shops. In this scenario, digital devices become as affordable as televisions, laptops and cameras are today. Thus, digital technologies help lower-class people experience virtual but vivid and immersive wildlife tourism experiences. For instance, they can experience swimming with whale sharks in the ocean, walking with elephants in the jungle or holding koalas in their hands. These technologies can even create experiences that could not exist in reality, so consumers can experience imaginary activities: swimming with whales under a starry sky, flying with an eagle or interacting with some extinct species. Despite the strengths of digital experiences, they have not totally substituted the real and embodied wildlife tourism experience. On-site travel becomes a symbol of social prestige and privilege under the condition of digital experiences becoming highly accessible in daily life.

Even though green growth through decoupling is the main policy framework proposed by the UN, there are many critiques of this concept. Decoupling is claimed to ignore the limits to the extraction of natural resources, and some developed countries achieve decoupling by extracting natural resources from other countries (Wiedmann *et al.*, 2015). Moreover, the ideas of decoupling attempt to achieve sustainability under the given political economic framework of neoliberalism, which means a capitalist economy that is 'internalizing the environmental impacts of growth to an unprecedented degree' (Fletcher & Rammelt, 2016: 4); it is thus argued that the green growth path to a sustainable future is a mere fantasy (Fletcher & Rammelt, 2016).

Scenario 2: Degrowth to a sustainable future

This scenario follows another path towards a sustainable future: degrowth. Degrowth is defined as the 'equitable downscaling of production and consumption that increases human well-being and enhances ecological conditions at the local and global level, in the short and long term' (Schneider *et al.*, 2010: 512). On this path, reducing production and consumption is the essential approach to sustainability (Schneider *et al.*, 2010).

According to the review of Cosme *et al.* (2017), a degrowth future entails fundamental social changes in production, consumption, environmental conservation, wealth distribution and political forms. With all of the global political entities conscious of the urgent need to launch social transformations, governments reduce large-scale and resource-intensive production and impose heavy taxes on resource extraction and environmental pollution. Production systems prefer the use of local resources and distribute products locally. Social actors change their lifestyles by reducing consumption, generating less waste and caring for the environment. Infrastructure construction changes from being rapid and based on private car use to being slow and supporting environmentally sustainable means of transport. Laws are established to reduce the gap between rich and poor. Community-based production is encouraged and provides equal job opportunities for community members. Basically, the degrowth path promises a localised and self-sufficient lifestyle. This social imagination is more likely to be fulfilled in a democratic and equal political environment.

In this path, fast and long-haul tourism is not encouraged in the ideological, cultural and material dimensions. In the 2014 movie *Interstellar*, directed by Christopher Nolan, future people on Earth face energy shortages and extreme weather conditions. The social ideology encourages people to focus on their daily lives rather than outer space, to become farmers rather than engineers or astronauts. The school curriculum brainwashes people to believe that the Moon landing was a hoax and restrains people's curiosity about faraway places. Similarly, in a degrowth future, social-cultural discourse discourages the desire for consumption based on high energy use, including fast, long-distance travel. Moreover, the infrastructure does not support such means of travel. The general reduction in demand causes a decline in tourist destinations.

In terms of wildlife tourism, only limited non-consumptive wildlife tourism products remain, most of which are located in remote protected areas. Because the social infrastructure and culture do not support long-haul travel, only a small percentage of tourists with a special interest in wildlife will seek opportunities to encounter animals in protected areas. Because of the equality of the social structure, there is no super-rich group who can afford high-priced fast travel. Therefore, almost all non-consumptive wildlife tourists travel to their destinations using slow transport, such as walking, cycling or using public transport. This equates to a kind of ascetic and slow tourism. Facilities aimed at tourists are underdeveloped, leaving the ascetic tourists highly dependent on local restaurants, accommodation and transportation. In protected areas, only a few corporations are providing wildlife tourism products and services. Consumptive wildlife tourism, such as trophy hunting, is prohibited because of the harm to animals. In any case, few people can afford a trophy hunting tour due to the socioeconomic situation.

Most people develop an interest in their surrounding environment. People engage in recreation in their own gardens, urban parks, zoos and aquariums. Zoos and aquariums continue to play important roles as species shelters, research institutions and education centres. However, they no longer import exotic species and are tasked only with raising local species. Urban residents can enjoy recreation and education through urban zoos and aquariums. Urban parks become vital again. Because of ecological restoration, more wildlife returns to urban areas. The BBC documentary *Cities: Nature's New Wild* shows the possibility of co-existence between wild animals and urban residents. For instance, the urban waterways of New York, one of the largest cities in the world, are again becoming home to humpback whales; in the city of Adelaide, people can watch flying foxes in many city parks; in Thailand's Lopburi City, long-tailed monkeys live alongside the residents. Given the restoration of urban ecology, urban parks become new leisure places for people.

Although proponents of degrowth present a picture of an ideal, peaceful and convivial future, its feasibility is still in doubt (Schwartzman, 2012). Degrowth has drawn much attention from academia but remains a marginalised viewpoint and lacks policy support (Sandberg et al., 2019). In the current capitalist political economy, it is difficult to reverse people's 'growth addiction' (van Griethuysen, 2010) and voluntarily reduce the size of the economic system. There is still severe competition between global economic entities, making it insecure for a country to be the first to adopt a degrowth path. The history of global negotiations related to climate change reveals that it is difficult to achieve an international consensus on plans for future GHG emissions; it is even more fantastical to hope that some countries will voluntarily reduce their national levels of production and consumption.

Scenario 3: Dark growth and an unsustainable future

John Urry (2016) is correct to posit a high possibility of failure to solve the climate change problems and continuing the same growth patterns to the point of catastrophe. I name this scenario 'dark growth' for its hint at an unsustainable future.

Dark growth involves energy use, production and consumption continuing at the same or a higher level than today. No new clean technologies are invented. In consequence, energy resources are exhausted, the environment is severely damaged and global temperatures keep rising. The negative impacts of climate change on the social and economic spheres begin to emerge. According to Carleton and Hsiang's (2016) review, anthropogenic climate change will directly increase heat-related mortality rates and indirectly cause mortality by inducing tropical cyclones, floods and droughts. Climate change will also influence the economy by impacting on agriculture, labour supply and productivity, energy supply and demand, and trade (Carleton & Hsiang, 2016). In the social aspects, global warming might influence females' incomes and mortality rates and increase interpersonal and intergroup violence (Carleton & Hsiang, 2016). Global heating may also increase migration and change population structures and growth.

As Urry (2016) points out, a dark growth future is a 'new medieval' world. Different interest groups scramble for the exhausted energy, food, water and natural resources and living space. Within and between countries, people do not avoid resorting to violence to solve conflicts. Inequality in all respects is greater, with class differentiations and relevant spatial segregations becoming more apparent and severe. Cities are gated from the 'wild zone' outside (Urry, 2016). Rich people live in urban spaces with sufficient food and water, armed protection and good medical care. In contrast, poor people live in the wild zone with poor living conditions, shortages of food and water and few job opportunities. Many Hollywood movies have imagined the unequal living conditions. For example, in the 2013 movie *Snowpiercer*, directed by Bong Joon Ho, all of the human survivors of a catastrophic global cooling experiment live on a train, with lower-class and upper-class people living in different coaches with different living standards. In the 2013 movie *Elysium*, directed by Neill Blomkamp, the powerful and wealthy class live on a space station with access to resources and militarised protection, and the poorer classes live on the resource-depleted Earth. On Earth, severe conflicts take place between the fortressed cities and the wild zones.

This spatial mode does not consider wildlife conservation. Human activities, such as overexploitation, agriculture, urban development, invasion and disease, pollution, system modification and climate change represent huge threats to biodiversity (Maxwell et al., 2016). Thomas et al. (2004) predict that on the basis of mid-range climate-warming scenarios for 2050, 15–37% of species will become extinct. Even though recent levels of climate change accelerate extinction risks (Urban, 2015), wild animals remain in protected areas, even in the dark growth future. However, the challenges of protecting these species are unprecedented, not only from ecological degradation but also from social conflict. Lacking food and water, poor people living in wild zones enter protected areas to hunt endangered animals. In response, the protected areas are secured by arms. This is a form of militarised conservation that has already emerged in contemporary Africa (Duffy et al., 2019). For instance, the documentary *Virunga* shows conflicts between conservationists and poachers in Congo's Virunga National Park. Warlords, terrorists and other poachers hunt gorillas for money. These situations are common in the dark growth future, with governments using military forces to protect endangered species. The macro-spatial model of a dark growth future is therefore *fortressed cities – wild zones – militarised nature reserves*.

In this social and spatial condition, there are also non-consumptive, low-consumptive and consumptive wildlife tourism. First, because energy resources are depleted and expensive, most people living in cities and wild zones cannot afford long-distance travel. Only the super-rich can visit these nature fortresses, which are distant from the cities, using aircraft or armed land transport to reach nature reserves. In the protected areas, these wealthy tourists can watch and photograph wild animals. Luxurious tourism services are provided in the reserves for the rich and powerful. Second, there is the possibility of hunting wild animals under scientific control in the nature reserves for those willing to pay a high price. Wildlife habitats are fragmented and fenced for biological security. The conservation organisations and tourism companies may choose old, sick or disabled animals for trophy hunting. Third, urban zoos still exist for protecting and researching endangered species. Urban residents can still enjoy leisure time in zoos but, due to the large loss of biodiversity, very few animals are kept there.

People living in wild zones have few opportunities for wildlife tourism. For them, an encounter with wild animals is invaluable because it provides food. However, due to the militarisation of conservation, they seldom have a chance to hunt wild animals in nature reserves.

Conclusion

The future of wildlife tourism cannot be isolated from macro social developments. Urry's (2015, 2016) work inspires this chapter to develop three scenarios for future social developments: green growth, degrowth and dark growth. Green growth relies on marketing and technological innovations to guarantee a sustainable future, in which tourists from the upper-middle classes can still experience consumptive, low-consumptive and non-consumptive wildlife tourism. Meanwhile, the lower classes also have opportunities to encounter wildlife in virtual reality using highly developed technologies. Degrowth proponents claim the need to reduce social production and consumption to attain a sustainable living environment. In a degrowth future, people develop new interests in wildlife in nearby green spaces, with only a few tourists with special interests keen to engage in slow and ascetic means of travel to visit remote areas and encounter exotic animals. A dark growth future is one in which global societies continue in their existing growth patterns and face the dangers of civilisational collapse caused by climate change. The upper and middle classes live in fortressed cities and the lower classes live in wild zones outside of these cities. Most of the remaining wildlife inhabit militarised nature reserves. Only the super-rich and powerful have the opportunity to experience wildlife tourism in nature reserves. For people in the wild zones, wildlife is valued as food rather than as a tourist attraction.

Green growth is the most preferred future for its promise to achieve sustainability without reducing the current quality of life. However, our social systems have co-evolved with the development of market capitalism to internalise economic growth and the extraction of natural resources (Büchs & Koch, 2019; Fletcher & Rammelt, 2016). Green growth through decoupling may therefore be a mere fantasy (Fletcher & Rammelt, 2016). In a global competitive structure, degrowth is also difficult to achieve. Social patterns are locked in, and this reduces the openness of the future. Modern human civilisation may be facing collapse due to the competitive and predatory exploitation of natural resources. Therefore, Urry (2016) is correct to posit dark growth as the most possible global future. In this catastrophic scenario, wildlife tourism is not important for most people, and tourism and travel become limited and privileged leisure activities.

This study has many implications for future research and industrial practices. First, existing studies mainly consider the benefits that wildlife tourism can bring to biodiversity conservation on a local scale. Future research needs to examine more about the costs that non-consumptive wildlife tourism generates on carbon emission, which may exacerbate global warming and biodiversity loss on a global scale (Thomas et al., 2014). Second, climate change intensifies the instability of the natural environment and may cause destructive damage to wildlife, for instance, the 2019 Australian wildfires caused 1 billion wildlife deaths. Researchers should reconsider the importance of zoos in species conservation and environment education, because zoos can provide shelter for endangered species in the wild, and zoo tourism contributes to mitigating global warming by shortening travel length compared to non-consumptive wildlife tourism. Third, the use of digital technologies in tourism is an emerging trend. For example, for the COVID-19 epidemic, many Chinese urban zoos provide live experiences to the public who have to stay at home. The technology of live broadcasting is a promising way to experience wildlife in remote protected areas. If the industry can invent more immersive and interactive live wildlife tourism activities, it will both mitigate global warming and raise funds for wildlife conservation.

Acknowledgements

This research is supported by National Natural Science Foundation of China (No. 41901161) and Natural Science Foundation of Guangdong Province (No. 2018A030310252). I would like to express my gratitude to Professor Honggang Xu for inspiring discussions.

References

Alves, R.R.N. and Lechner, W. (2019) Wildlife attractions: Zoos and aquariums. In R.R.N. Alves and U.P. Albuquerque (eds) *Ethnozoology: Animals in Our Lives* (pp. 351–361). London: Academic Press.

Amer, M., Daim, T.U. and Jetter, A. (2013) A review of scenario planning. *Futures* 46, 23–40.

Barua, M. (2017) Nonhuman labour, encounter value, spectacular accumulation: The geographies of a lively commodity. *Transactions of the Institute of British Geographers* 42 (2), 274–288.

Bell, W. (2003) *Foundations of Future Studies*. Piscataway, NJ: Transaction Publishers.

Bina, O., Mateus, S., Pereira, L. and Caffa, A. (2017) The future imagined: Exploring fiction as a means of reflecting on today's Grand Societal Challenges and tomorrow's options. *Futures* 86, 166–184.

Bostock, S.S.C. (2003) *Zoos and Animal Rights*. London/New York: Routledge.

Bostrom, N. and Cirkovic, M.M. (eds) (2011) *Global Catastrophic Risks*. Oxford: Oxford University Press.

Büchs, M. and Koch, M. (2019) Challenges for the degrowth transition: The debate about wellbeing. *Futures* 105, 155–165.

Carleton, T.A. and Hsiang, S.M. (2016) Social and economic impacts of climate. *Science* 353 (6304), aad9837.

Castley, J.G. (2016) Wildlife tourism. In J. Jafari and H. Xiao (eds) *Encyclopedia of Tourism* (pp. 1020–1021). Cham: Springer International Publishing.

Ceballos, G., Ehrlich, P.R., Barnosky, A.D., García, A., Pringle, R.M. and Palmer, T.M. (2015) Accelerated modern human–induced species losses: Entering the sixth mass extinction. *Science Advances* 1 (5), e1400253.

Cosme, I., Santos, R. and O'Neill, D.W. (2017) Assessing the degrowth discourse: A review and analysis of academic degrowth policy proposals. *Journal of Cleaner Production* 149, 321–334.

Crutzen, P.J. (2002) Geology of mankind. *Nature* 415, 23.

Currie, A.M. and Ó hÉigeartaigh, S. (2018) Working together to face humanity's greatest threats: Introduction to the future of research on catastrophic and existential risk. *Futures* 102, 1–5.

Di Minin, E., Leader-Williams, N. and Bradshaw, C.J. (2016) Banning trophy hunting will exacerbate biodiversity loss. *Trends in Ecology & Evolution* 31 (2), 99–102.

Duffus, D.A. and Dearden, P. (1990) Non-consumptive wildlife-oriented recreation: A conceptual framework. *Biological Conservation* 53 (3), 213–231.

Duffy, R., Massé, F., Smidt, E., Marijnen, E., Büscher, B., Verweijen, J., Ramutsindela, M., Simlai, T., Joanny, L. and Lunstrum, E. (2019) Why we must question the militarisation of conservation. *Biological Conservation* 232, 66–73.

Dybsand, H.N.H. (2020) In the absence of a main attraction – Perspectives from polar bear watching tourism participants. *Tourism Management* 79, 104097.

Fletcher, R. and Rammelt, C. (2016) Decoupling: A key fantasy of the post-2015 sustainable development agenda. *Globalizations* 14 (3), 450–467.

Hall, C.M., Amelung, B., Cohen, S., Eijgelaar, E., Gössling, S., Higham, J., Leemans, R., Peeters, P., Ram, Y. and Scott, D. (2015) On climate change skepticism and denial in tourism. *Journal of Sustainable Tourism* 23 (1), 4–25.

Hill, J., Curtin, S. and Gough, G. (2014) Understanding tourist encounters with nature: A thematic framework. *Tourism Geographies* 16 (1), 68–87.

IPCC (2013) Summary for policymakers. In T.F. Stocker, D. Qin, G.-K. Plattner, M. Tignor, S.K. Allen, J. Boschung, A. Nauels, Y. Xia, V. Bex and P.M. Midgley (eds) *Climate Change 2013: The Physical Science Basis* (pp. 3–29). Cambridge: Cambridge University Press.

IPCC (2014) *Climate Change 2014: Synthesis Report. Contribution of Working Groups I, II and III to the Fifth Assessment Report of the Intergovernmental Panel on Climate Change* (Core writing team: R.K. Pachauri and L.A. Meyer [eds]). Geneva: IPCC.

Kallis, G. (2011) In defense of degrowth. *Ecological Economics* 70 (5), 873–880.

Lenzen, M., Sun, Y.Y., Faturay, F., Ting, Y.P., Geschke, A. and Malik, A. (2018) The carbon footprint of global tourism. *Nature Climate Change* 8 (6), 522–528.

Maxwell, S.L., Fuller, R.A., Brooks, T.M. and Watson, J.E. (2016) Biodiversity: The ravages of guns, nets and bulldozers. *Nature News* 536 (7615), 143–145.

Mitas, O. and Bastiaansen, M. (2018) Novelty: A mechanism of tourists' enjoyment. *Annals of Tourism Research* 72, 98–108.

Pratt, S. and Suntikul, W. (2015) Can marine wildlife tourism provide an 'edutaining' experience? *Journal of Travel & Tourism Marketing* 33 (6), 867–884.

Sandberg, M., Klockars, K. and Wilén, K. (2019) Green growth or degrowth? Assessing the normative justifications for environmental sustainability and economic growth through critical social theory. *Journal of Cleaner Production* 206, 133–141.

Schneider, F., Kallis, G. and Martinez-Alier, J. (2010) Crisis or opportunity? Economic degrowth for social equity and ecological sustainability. Introduction to this special issue. *Journal of Cleaner Production* 18 (6), 511–518.

Schwartzman, D. (2012) A critique of degrowth and its politics. *Capitalism Nature Socialism* 23 (1), 119–125.

Shani, A. and Arad, B. (2014) Climate change and tourism: Time for environmental skepticism. *Tourism Management* 44, 82–85.

Steffen, W., Broadgate, W., Deutsch, L., Gaffney, O. and Ludwig, C. (2015) The trajectory of the Anthropocene: The great acceleration. *The Anthropocene Review* 2 (1), 81–98.

Thomas, C.D., Cameron, A., Green, R.E., Bakkenes, M., Beaumont, L.J., Collingham, Y.C., Erasmus, B.F.N., Ferreira de Siqueira, M., Grainger, A., Hannah, L., Hughes, L., Huntley, B., van Jaarsveld, A.S., Midgley, G.F., Miles, L., Ortega-Huerta, M.A., Townsend Peterson, A., Phillips, O.L. and Williams, S.E. (2004) Extinction risk from climate change. *Nature* 427 (6970), 145–148.

Tollefson, J. (2019) One million species face extinction. *Nature* 569, 171.

United Nations Environment Programme (UNEP) (2011) Decoupling Natural Resource Use and Environmental Impacts from Economic Growth. A report of the working group on decoupling to the international resource panel.

Urban, M.C. (2015) Accelerating extinction risk from climate change. *Science* 348 (6234), 571–573.

Urry, J. (2013) *Societies Beyond Oil: Oil Dregs and Social Futures*. London/New York: Zed Books Ltd.

Urry, J. (2015) Climate change and society. In J. Michie and C.L. Cooper (eds) *Why the Social Sciences Matter* (pp. 45–59). London: Palgrave Macmillan.

Urry, J. (2016) *What is the Future?* Cambridge: Polity Press.

Van Griethuysen, P. (2010) Why are we growth-addicted? The hard way towards degrowth in the involutionary western development path. *Journal of Cleaner Production* 18 (6), 590–595.

Wiedmann, T.O., Schandl, H., Lenzen, M., Moran, D., Suh, S., West, J. and Kanemoto, K. (2015) The material footprint of nations. *Proceedings of the National Academy of Sciences* 112 (20), 6271–6276.

3 Rabbits in the Wild: Close Encounters on an Equal Footing?

Rie Usui and Carolin Funck

Rabbits are a familiar animal to many people around the world; they are widespread and are often kept as household pets. Despite their ubiquity, their popularity increased when hundreds of feral rabbits exhibiting tame behaviour occupied an island landscape – Ōkunoshima Island (0.7 km^2) – located in Hiroshima Prefecture, Japan (Figure 3.1). Yet, their recent gain in popularity among tourists has resulted in uncontrolled tourist–rabbit encounters, owing to the rabbits' presence on the island as a feral animal that is positioned somewhere on the wild–domestic continuum (Griffiths et al., 2000). Feral animals are initially domesticated (Notzke, 2016) but at one point escaped from the captive environment (Brehm, 2000) and proliferated in the wild (Berger, 1986 cited in Bough, 2016). They challenge our spatial ordering of animals and, in the context of wildlife tourism, they blur the notion of wild (Bulbeck, 2005; Notzke, 2016). Ferality poses ethical questions concerning the treatment of a feral species as a major target in tourism destinations. How should we deal with wildlife tourism sites where the subject animals are considered 'out of place'? To suggest a future path, this study develops two scenarios using Ōkunoshima Island as a case study.

The remainder of this chapter is organised into five sections. The first section provides an overview of Ōkunoshima Island to illustrate the context of the island, including a brief history. The second section describes this study's methodology based on a normative scenario approach. The third section evaluates the current situation on Ōkunoshima Island, and the fourth section presents two scenarios for Ōkunoshima Island to help us imagine the future of the island: one situated within the sustainable tourism path, and the other uses an ecofeminism approach. We address the following question: What would the rabbits' future be like if sustainable tourism and ecofeminism approaches were adopted on Ōkunoshima Island? The final section presents the authors' conclusion.

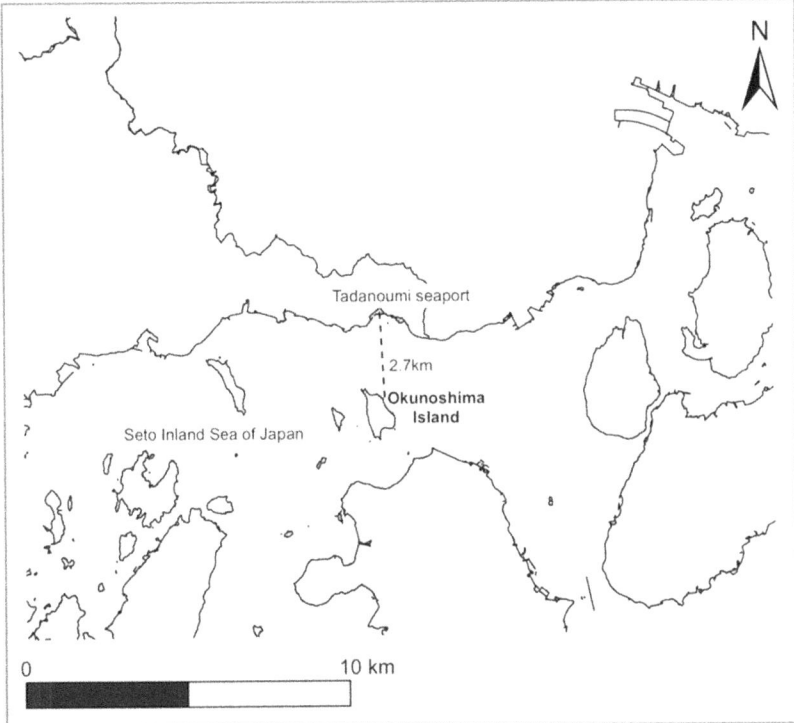

Figure 3.1 The location of Ōkunoshima Island

An Overview of Ōkunoshima Island

Ōkunoshima Island underwent a dark period during the Second World War. Unlike the recent image of the island associated with its common name 'Paradise of Rabbits' or 'Rabbit Island', it was once described as 'the island that was removed from the map'. From 1929 to 1945, Ōkunoshima Island was designated a secret military base for producing chemical weapons that were used by the Japanese Army (Sato, 2010; Takeo, 2010) and distributed throughout Japan and parts of China (Sato, 2010). According to Takeo (2010), 6500 local residents from the neighbouring communities were employed to manufacture chemical weapons under a strict ordinance that prohibited them from discussing the island and its military mission.

Ōkunoshima Island was registered as a National Vacation Village in 1960, and a visitor centre was established in 2003 to promote the island as a tourism destination, where tourists could learn about the island's natural features (Takeo, 2010). The island has various recreational facilities, including a hotel, tennis courts and beaches that form part of the

26 Part 1: Paths Towards the Futures of Wildlife Tourism

Setonaikai National Park. Even after the island took on its new role as a recreational destination, facilities used during the Second World War remained on the island. Ruins such as chemical weapon storages and a power plant can be found on the island today (Figure 3.2), providing visitors with the opportunity to learn about Ōkunoshima Island's Second World War history.

Until 2015, Ōkunoshima Island was only known to a limited number of local tourists, but from that year, it began attracting an unprecedented number of domestic and international tourists. In 2016, the annual tourist number reached its peak, exceeding 223,000 visitors (Kunoshima Co.

Figure 3.2 A power plant ruin (Photo by Rie Usui)

Ltd., 2019). The increasing rate of international tourists has brought significant changes to the visitor profile of the island. The international tourist numbers, which were less than 1000 per year prior to the tourism boom, increased dramatically to 19,000 in 2016 (Takehara City, 2017). Hence, what was once a domestic tourist destination is now an international tourist destination (Usui *et al.*, 2017).

Owing to the history of Ōkunoshima Island and its chemical weapons production, there has been speculation that the rabbits were used for testing the chemical weapons or that they were brought to the island to examine whether gas residue remained on the island after the Second World War (personal communication). However, how the rabbits ended up on Ōkunoshima Island remains unclear. Another common explanation is that school children brought a couple of rabbits to the island, left them there and they proliferated over time (personal communication). A local man from Tadanoumi town, the neighbouring town from where the ferry to Ōkunoshima Island departs, also claimed that he released his pet rabbits onto the island (personal communication). Regardless of the true origin of the rabbits' existence on the island, they are likely to have been brought by humans some time during or after the Second World War.

Many tourists on the ferry ride to the island can be observed carrying food for the rabbits. Additionally, some tourists leave extra food for the rabbits when they depart the island. Due to feeding by tourists, combined with the fact that they have no major predator, the rabbit population on the island has grown exponentially, and several problems associated with this population explosion have been reported, such as rabbits' health issues and damage to the island's vegetation.

A further complication is that the rabbits on Ōkunoshima Island are not native to Japan. The species of rabbits that inhabit the island are European rabbits (*Oryctolagus cuniculus*), listed as an invasive alien species by the Japanese Ministry of the Environment (2016). They originate from the Iberian Peninsula and were supposedly brought to Japan sometime during the 18th century (Sakurai, 2016). Therefore, under the Japanese Wildlife Protection and Hunting Law, Ōkunoshima Island's rabbits are a species that needs to be eradicated.

Public opinion on who is responsible for managing the rabbits and how this management should be implemented is divided. Signs on the ferry inform tourists of the rules on how to interact with the rabbits; however, thus far, tourist–rabbit interactions have not been regulated officially by any management entity. The island's only hotel used to sell rabbit food and set up a water station for the rabbits; however, rabbit food is no longer sold on Ōkunoshima Island. Instead, the ferry ticketing office at Tadanoumi port sells small bags of pellets (Figure 3.3), with the amount of pellets in the bag adjusted by season. For instance, there are more pellets per bag during the winter when tourist numbers are smaller than during the spring and summer seasons. Although the ferry ticketing

Figure 3.3 'Bunny food' sold at the ticketing office (Photo by Rie Usui)

office provides the rabbit food, the national park prohibits feeding the rabbits. Thus, the nature interpreters who work for the national park discourage tourists from feeding the rabbits, but they have no power to stop tourists from doing so. Consequently, tourists visit and interact freely with the island's rabbits. Sometimes, people abandon their pet rabbits on the island.

In response to the absence of an official management entity, some concerned tourists have formed a self-organised volunteer activity called *usakatsu* (Usui & Funck, 2018). It is an abbreviated Japanese name, meaning 'activities for rabbits'. As the name suggests, volunteer tourists engage in activities that they consider necessary to protect the rabbits on the island, including providing fresh water; supplementary provision of food, especially during a low-tourist season; cleaning up discarded food; and monitoring general visitors. Some volunteer tourists also try to raise visitors' awareness by teaching them how to handle the rabbits appropriately and educating them on what not to feed the rabbits. Many of them share their activities on their blogs. While these volunteer tourists fulfil the roles that are otherwise absent on the island (i.e. management roles), their abilities are limited to mitigating the rabbits' present situation. Moreover, one of the *usakatsu* visitors pointed out that the motivations and purposes of the *usakatsu* visitors vary, indicating that their grassroots activities are unlikely to solve the fundamental issue of the island.

Methodology

To imagine the future of Ōkunoshima Island, we adopted a scenario planning approach, which has been well established and applied widely in many cases such as water resource management (e.g. Schneider & Rist,

2014) and landscape planning (e.g. Nassauer & Corry, 2004). In tourism, the scenario planning approach can be a useful means to challenge the traditional way of thinking and to help us envision alternatives (Yeoman *et al.*, 2011, 2015). Specifically, we used a normative scenario approach (van Notten *et al.*, 2003) to examine the future of Ōkunoshima Island's wildlife tourism landscape. The normative scenario begins by setting a desirable future goal (Morioka *et al.*, 2006; van Notten *et al.*, 2003) and working backwards to achieve such a societal goal (Nassauer & Corry, 2004).

First, the scenario development process requires collecting existing data (Nassauer & Corry, 2004). This study presents data based on our observations and interviews conducted during multiple site visits between July 2016 and December 2018. The interview respondents include tourists (Japanese and international), a nature interpreter, an owner of the ferry ticket company and a local volunteer guide who provides a history tour of the island. We use Bertella's (2018) reflections on the sustainability of wildlife tourism and possible alternative approaches as the point of departure to develop scenarios.

Evaluating the Current Situation of Ōkunoshima Island

As the first step in the normative scenario approach, we assessed the current situation of Ōkunoshima Island (Table 3.1), and we used Shani and Pizam's (2008) ethical framework to evaluate the tourist–rabbit interactions on the island. Their model proposes three principal components of the ethical guidelines of animal-based tourist attractions: entertainment, education and animal welfare.

Table 3.1 Evaluation of Ōkunoshima Island's situation based on three ethical components by Shani and Pizam (2008)

Component	Evaluation of the current situation
Entertainment	• The site offers a relatively high degree of entertainment to tourists because they have the freedom to engage in any kind of interactions with the rabbits (e.g. viewing, taking photos, feeding, touching and holding). • There is currently no responsible organisation or personnel to manage the tourist–rabbit interactions.
Education	• The absence of an institutionalised educational policy could create a situation wherein the wrong information is provided, or various stakeholders inform tourists differently.
Animal welfare	• The fact that there are close to a thousand rabbits on the island indicates that the living conditions of the rabbits are far from their natural state, meaning that the rabbits compete for limited resources (e.g. food and space); thus, the rate of intra-species conflict is likely to increase, resulting in possible injuries and poor health. • Without *usakatsu* visitors, the rabbits' access to water would be highly limited. • Some rabbits exhibit abnormal behaviours or show signs of poor health. • Occasionally, dead rabbits are spotted along the trails.

The first component, entertainment, concerns the role of wildlife tourism destinations in facilitating responsible encounters of tourists with animals, essentially requiring the balancing of the needs of the visitors and the animals. For instance, in captive settings, highly visible animals can increase visitor satisfaction; however, it should not sacrifice the welfare of the animals. A responsible encounter prevents or minimises the potential negative impacts that could result from tourist–animal interactions. In non-captive settings, it is likely that there are closer interactions compared to captive settings, and responsible encounters should consider the consequences of such behaviours on the animals' health. Currently, injured rabbits are left alone, even if tourism-related activities such as tourists dropping the rabbits on the ground or cyclists running over the rabbits caused the injury.

Tourists who visit Ōkunoshima can engage in any interactions they wish to have with the rabbits, ranging from non-physical interactions (e.g. viewing and taking photos) to physical interactions that involve direct contact with the rabbits (e.g. feeding, touching and holding). The fact that there is no strict regulation of tourists' behaviour indicates that the site offers a relatively high degree of entertainment. Additionally, the absence of a responsible organisation or personnel to manage the tourist–rabbit interactions suggests that the needs of the rabbits are being overlooked, supporting the fact that the site is solely for the entertainment of tourists.

The second component concerns the educational aspect of tourist–rabbit interactions. Education can be offered in the form of an information booklet or lecture, or it can be combined with entertainment activities, for instance, by engaging tourists in a quiz. The goal is to raise awareness of the environmental issues concerning the animals the tourists have come to see. At Ōkunoshima Island, different stakeholders (e.g. ferry company and *usakatsu* visitors) independently inform tourists about appropriate interactions with the rabbits. For instance, some *usakatsu* visitors tell tourists not to hold the rabbits because the rabbits are fragile and such interactions may result in the rabbits being dropped, and they inform the tourists about food that is inappropriate for the rabbits. The nature interpreter also provides information regarding the rabbit population and basic knowledge about the rabbits. Nevertheless, institutionalised education is lacking.

The third ethical component concerns the animal welfare principle, which asks whether animals are given enough space; whether they can exhibit their normal behaviour; whether they have a place to hide; and whether they are free from pain, suffering, thirst and starvation. In Ōkunoshima, the number of rabbits on the island is said to have exceeded the carrying capacity of the island due to feeding by tourists, implying that their 'normal' behaviours may have been altered and intra-species aggression may have increased. Access to water, a requirement for the animals' welfare, is ensured by the *usakatsu* visitors. Some rabbits exhibit abnormal behaviours (e.g. sneezing repetitively and limping) or show signs of poor health (e.g. patchy hair, bite marks, tumours and split ears). Furthermore, tourist numbers on the island vary greatly between holidays; from spring to autumn, there are up to 2000 visitors per day, but visitors are few in the winter season. This discrepancy creates an uneven food supply that might be detrimental to the rabbits' health. The open environmental condition on Ōkunoshima Island somewhat allows the rabbits to make a choice when tourists interact with them. For instance, when tourists feed the rabbits, an important component of the tourists' entertainment, the rabbits can choose to receive food from tourists, ignore the tourists or retreat if they are bothered by them. Still, the survival of Ōkunoshima Island's rabbits largely depends on tourists because they essentially provide food to them.

To improve the current situation of Ōkunoshima Island, it is necessary to have an organisation or a person who is in charge of managing tourist–rabbit interactions. It is likely that the lack of such an entity means the rabbits' overall well-being will not be satisfied, which will in turn negatively influence tourists' satisfaction. By recognising such a change as a bottom-line requirement, let us imagine two possible future paths for Ōkunoshima Island.

Scenario 1: Sustainable Wildlife Tourism as a Mainstream Approach

The notion of sustainability has been widely accepted and practiced in tourism in general, and wildlife tourism is no exception (e.g. Higham & Bejder, 2008; Kontogeorgopoulos, 1999; Moore & Rodger, 2010; Rodger *et al.*, 2007). While there have been numerous debates about the concept of sustainable tourism or sustainability (Bramwell *et al.*, 2016; Hardy *et al.*, 2002; Moscardo & Murphy, 2014; Saarinen, 2006; Zolfani *et al.*, 2015), the essential idea is 'to practice tourism that considers fully its current and future economic, social and environmental impacts, addressing the needs of visitors, the industry, the environment and host communities' (UNWTO, n.d.). Sustainability has been criticised as an anthropocentric concept that places humans as the central element in the world. Its primary interest is to serve the needs of humans (Bertella, 2018). In the context of sustainable wildlife tourism, sustainability is expected to conserve wild animals while generating economic revenue for the local communities (Newsome *et al.*, 2005). This scheme places value on the ecosystem as a whole. Hence, previous wildlife tourism studies have commonly measured the impact of tourism on ecology, the population and behaviours of animals, and the animal habitats for the purpose of conservation (e.g. Blumstein *et al.*, 2017; McDonald, 2009; Russon & Wallis, 2014; Smith *et al.*, 2008).

Based on the idea of sustainability, the future of Ōkunoshima Island would be one of an island without rabbits, as the rabbits are not native species to the island and severely affect the vegetation. Thus, the rabbits would either be eradicated or relocated elsewhere. A less radical approach would be to gradually reduce their population to prevent the detrimental impacts of the rabbits on the island's vegetation; however, ultimately, the population would be reduced to zero. Strictly prohibiting tourists from feeding the rabbits would be one way to reach this goal. However, it may only increase pressure on the vegetation and likely result in mass death of the rabbits immediately after such a solution is implemented.

While tourists in Japan are less sensitive to crowds in natural surroundings (Weaver, 2002) and setting limits on tourist numbers in national parks is not popular in Japan (Forbes, 2012), limiting tourist numbers would be another approach to solving issues such as overfeeding of rabbits on busy days, traffic congestion and negative tourist experiences because of overcrowding. Spaying most of the rabbits to control reproduction would also be a solution in line with sustainability, as was implemented to control stray cat populations elsewhere in Japan (Setouchi, 2017). However, spaying the rabbits to reduce their population could be problematic because there might still be possible damage to the local vegetation.

Scenario 2: Ecofeminism as an Alternative Approach

Ecofeminism bases its philosophical orientation on feminism and believes that there is a similar link between the domination of women by men and the domination of animals by humans (Fennell, 2015), thus rejecting a hierarchical structure that assumes human superiority over animals (Bertella, 2017). While both feminists and ecofeminists are concerned with the oppression and domination of subdominant beings, the latter extends their concern to nature (Warren, 1990). Oppressive attitudes and behaviours are essentially rooted in a dualistic notion, considering one superior over the other; for instance, culture over nature, male over female, and human over nature (Plumwood, 1993).

Unlike the sustainable tourism approach that places a higher value on the entire ecosystem (Fennell, 2014; Newsome *et al.*, 2005), ecofeminism denies humans' masculine relation to animals (Kheel, 2008) and considers animals as individual others that have emotions together with human individuals (Bertella, 2018). Thus, the ecofeminism perspective would attend to and care about the individual animals and their well-being, extending care and empathy to them (Yudina & Fennell, 2013). Care does not necessarily mean leaving the animals alone (Kheel, 2008), but denotes performing responsible acts towards animals (Bertella, 2018). Empathy is a prerequisite of the action for such caring behaviours, and thus contextualised care is required (Bertella, 2018).

What does it mean to adopt the ecofeminism notion to wildlife tourism? As Vance (1997: 69) argues, 'Any ecofeminist analysis of wilderness has to begin with a close examination of the very idea of wilderness', the term 'wildlife' needs careful consideration in the context of wildlife tourism. Based on Vance's (1997) claim, this very idea of 'wildlife' in wildlife tourism should be questioned because it is partially created by the masculinity dualistic notion, which ecofeminism rejects. In other words, 'wildlife' is a cultural product that humans have created; therefore, '(i)f we stop imagining that wilderness (wildlife in this case) equals nature, we can see and assess it as we would any other cultural manifestation' (Vance, 1997: 70).

Returning to the case of Ōkunoshima Island, as mentioned, the rabbits are alien species that have become wild. The idea of 'alien' species under ecofeminism thinking is also a social construction, indicating the dualistic – things that belong versus things that do not belong to a certain place. Borrowing Vance's (1997) example, the central issue from the feminist perspective is whether 'alien' species should exist at all, rather than to *where* they should or should not belong.

Under the ecofeminism approach, the rabbits on Ōkunoshima Island would be recognised as individuals whose intrinsic value matters. The outcome of this approach would also view the tourist–rabbit interactions differently. Instead of simply accepting the psychological benefit of humans' experience with nature within the realm of recreation, the ecofeminism perspective concerns those whose values are favoured (Vance, 1997). Hence, this approach requires one to think about what the rabbits want and consider adequately the context in which the tourist–rabbit interactions occur. For instance, asking questions such as, 'Do the rabbits want to be petted or held?' and 'Are they enjoying the interactions with tourists?' are a good starting point from which to examine Ōkunoshima Island's tourism through the eyes of ecofeminism.

A relevant and important aspect concerning the tourist–rabbit interactions at Ōkunoshima Island that should be discussed is the practice of feeding the rabbits. As mentioned earlier, the ecofeminism approach is concerned with the oppression of animals (Warren, 1990). The question then is, 'Is the constant feeding by tourists at the island an oppressive act?' Tuan (1988) claimed that the fundamental motivation behind feeding animals is a feeling of superiority and that feeding is a form of showing one's power. While no research investigates adequately why humans feed animals (Ishida, 2013), many feeding occurrences at Ōkunoshima Island can be seen as an oppressive act rather than as a purely altruistic one because some tourists are observed to use food as a bait to satisfy their needs (i.e. take photos). By adopting the ecofeminism approach, we are required to re-evaluate what seemed to be an altruistic act of feeding the rabbits.

Discussion

This chapter evaluated the concept of sustainable wildlife tourism and the ecofeminism approach regarding the rabbits on Ōkunoshima Island. Table 3.2 summarises the basic concepts and practical implications of the two scenarios. In this discussion, we highlight potential ethical dilemmas that could arise by adopting the sustainable tourism paradigm, and we argue that the sustainable tourism paradigm does not meet the challenges of today's wildlife tourism practice and that the ecofeminism approach can be an alternative future.

If the concept of sustainable wildlife tourism was applied to the Ōkunoshima Island situation, actions to achieve such a vision would raise the following ethical questions: Is it morally right to eradicate the rabbits for the sake of conserving the native plant species? What should be protected? Who is responsible for the situation and the actions taken? It is arguably true that humans had already modified the island's vegetation and that the wildlife that inhabited the island had adapted to the changing environment. While vegetation damage by the rabbits is one of the concerns raised, it is actually human activities that have severely affected vegetation in the Seto Inland Sea area over the centuries, and Ōkunoshima Island is no different from other islands in the region. Many of the islands in the Seto Inland Sea had at one time or another been extensively modified into agricultural farmlands. Therefore, we can assume that most of the vegetation on the island is secondary and that wildlife has adapted to an ecosystem that has been altered by humans. In other words, there is no true wilderness. From this viewpoint, the

Table 3.2 Two scenarios for Ōkunoshima Island: Sustainability and ecofeminism

Approach	Scenario 1: Sustainability	Scenario 2: Ecofeminism
Main principle	An anthropocentric view, which places a non-human animal's value on an overall ecosystem at a collective level to ultimately serve human interests.	Value is placed on an ethic of responsibility and care, with emphasis on compassion, equity and need (Fennell, 2015: 29).
Attitudes towards other-than-human beings	Human dominates over nature, thus nature can be utilised for human benefit.	Criticises human's domination over nature and denies dualistic notions such as nature vs. culture, and human vs. animals.
View of the rabbits as tourist–rabbit interactions	Economic resource.	Individuals with intrinsic value that have rich cognitive and emotional lives.
Practical implications	The population would be eradicated if the sustainability paradigm was strictly applied; otherwise, it would be relocated, as the rabbits are considered 'alien' species. Prohibition on feeding, controlling the tourist number or spaying the rabbits could be used to achieve the goal.	The idea of an 'alien' species would be questioned. One must question whose values are favoured. Questions such as 'Do the rabbits want to be petted or held?' and 'Are they enjoying the interactions with tourists?' are important concerns for ecofeminists.

vegetation to be protected under the claim that it is 'natural' is also not 'natural'. Therefore, a discussion about the wildlife and nature that should be protected under the sustainability paradigm is necessary.

The term 'sustainable wildlife tourism' is often used in the context where the state of the environment is assumed to be in its 'natural' state. This assumption is evident in that the majority of wildlife tourism research is limited to a situation where animals are considered ecologically significant beings (Usui & Funck, 2017), meaning they are regarded as a part of nature and their cultural significance is rarely considered. Consequently, implementing wildlife tourism management based on the sustainability concept could generate ethical issues regarding those animals that have adapted to living closely with humans. The place where tourist–animal interactions occur is not limited to the locations we typically expect, as in the case of Ōkunoshima Island's rabbits. Therefore, the rabbits on the island may be considered 'unnatural' because they are not native to the island. However, while the Japanese Wildlife Protection and Hunting Law lists the rabbits as an alien species, nature is not static, meaning it has always been and will always be changing.

It is important to highlight that the eradication or relocation of the rabbits under the sustainable tourism paradigm would also destroy the basis of wildlife tourism on Ōkunoshima Island. This would adversely affect the local community because tourism on the island has developed because of the rabbits, and it has undoubtedly brought some positive economic and social impacts to Ōkunoshima Island. For instance, a ticketing kiosk for the ferry service has transformed both its exterior and interior over time to offer a variety of services to tourists, and the ferry that connects to Ōmishima Island, one of the neighbouring islands, via Ōkunoshima Island has been able to continue in service due likely to the growth in the number of passengers taking the ferry to Ōkunoshima Island regularly throughout the year. Therefore, wildlife tourism on the island has empowered the local economy.

By contrast, the ecofeminism approach would help us carefully attend to the context within which human–animal interactions occur (Bertella, 2017), thereby largely differentiating itself from the sustainable tourism approach. Vance (1997: 71) emphasises the importance of attending to the historical context: 'An ecofeminist position, in contrast, would encompass a concern for biodiversity but would nevertheless seek to understand the context in which a loss of biodiversity has occurred before moving directly to remediation'. Given that the rabbits were brought to the island by humans and have been the subjects of human entertainment to some degree (e.g. tourists feeding, petting, holding and taking photos with them), from the ecofeminist perspective, not considering their well-being would be irresponsible. The tourism stakeholders have also benefited economically (e.g. the accommodation and transportation business) and experientially (e.g. tourists) because of the presence of rabbits on the island. After considering these facts, is it morally right to claim that rabbits are 'alien' species? The ecofeminism approach questioned this term: Is there such a thing as an 'alien' species? Often, such a socially constructed categorisation is at the centre of an ethical debate about why Ōkunoshima Island rabbits are as they are.

Conclusion

This case concerning the management of the rabbit population on Ōkunoshima Island has implications not only for the island's future but also for other wildlife tourism destinations, where opinions on how to handle similar situations in which animals are roaming free in their environment are often divided. As far as sustainability is concerned, the case of the rabbits on Ōkunoshima Island may not even be relevant. However, the ecofeminism approach, questioning the dualistic notions, allows us to move away from the argument about whether the rabbits are an 'alien' species. Adopting the ecofeminists' perspective would provide a positive future towards a 'Paradise *for* Rabbits' for the sake of rabbits, rather than a 'Paradise *of* Rabbits' for human interests.

Acknowledgements

We thank Kunoshima Co. Ltd. for sharing their valuable information. We also extend our acknowledgement to Dr Bertella for giving us the opportunity to contribute to this book.

Funding

The author (RU) was supported by the Hiroshima University TAOYAKA Program for creating a flexible, enduring, peaceful society, funded by the Program for Leading Graduate Schools, Ministry of Education, Culture, Sports, Science and Technology.

References

Berger, J. (1986) *Wild Horses of the Great Basin*. Sydney: University of Chicago Press.
Bertella, G. (2017) An eco-feminist perspective on the co-existence of different views of seals in leisure activities. *Annals of Leisure Research* 21 (3), 284–301.
Bertella, G. (2018) Sustainability in wildlife tourism: Challenging the assumptions and imagining the alternatives. *Tourism Review* 47 (1), 1–11.
Blumstein, D.T., Geffroy, B., Samia, D.S.M. and Bessa, E. (2017) *Ecotourism's Promise and Peril: A Biological Evaluation*. Cham: Springer.
Bough, J. (2016) Our stubborn prejudice about donkeys is shifting as they protect Australia's sheep from wild dogs. *Australian Zoologist* 38 (1), 17–25.
Bramwell, B., Higham, J., Lane, B. and Miller, G. (2016) Twenty-five years of sustainable tourism and the *Journal of Sustainable Tourism*: Looking back and moving forward. *Journal of Sustainable Tourism* 25 (1), 1–11.
Brehm, C.E. (2000) Examination of the Bureau of Land Management's implementation of the Wild, Free-Roaming Horse and Burro Act at Red Rock National Conservation

Area. UNLV Theses, Dissertations, Professional Papers, and Capstones. 204, 1–33. See https://digitalscholarship.unlv.edu/thesesdissertations/204 (accessed 5 September 2020).

Bulbeck, C. (2005) Loving knowing. In C. Bulbeck (ed.) *Facing the Wild: Ecotourism, Conservation and Animal Encounters* (pp. 155–180). New York: Routledge.

Fennell, D. (2014) Exploring the boundaries of a new moral order for tourism's global code of ethics: An opinion piece on the position of animals in the tourism industry. *Journal of Sustainable Tourism* 22 (7), 983–996.

Fennell, D. (2015) The status of animal ethics research in tourism: A review of theory. In K. Markwell (ed.) *Animals and Tourism: Understanding Diverse Relationships* (pp. 27–43). Bristol: Channel View Publications.

Forbes, G. (2012) Yakushima: Balancing long-term environmental sustainability and economic opportunity. *Kagoshima Immaculate Heart College Bulletin* 42, 35–49.

Griffiths, H., Poulter, I. and Sibley, D. (2000) Feral cats in the city. In C. Philo and C. Wilbert (eds) *Animal Spaces, Beastly Places: New Geographies of Human–Animal Relations* (pp. 56–70). New York: Routledge.

Hardy, A., Beeton, R.J.S. and Pearson, L. (2002) Sustainable tourism: An overview of the concept and its position in relation to conceptualisations of tourism. *Journal of Sustainable Tourism* 10 (6), 475–496.

Higham, J.E.S. and Bejder, L. (2008) Managing wildlife-based tourism: Edging slowly towards sustainability? *Current Issues in Tourism* 11 (1), 75–83.

Ishida, O. (2013) 'Fureai' to osewa ['Interactions' and care]. In O. Ishida, S. Hamano, M. Hanazono and A. Setoguchi (eds) *Nihon no doubutsukan: hito to doubutsu no kankeishi [Japanese Perceptions of Animals: A History of Human–Animal Relationship]* (pp. 209–225). Tokyo: University of Tokyo Press.

Kheel, M. (2008) *Nature Ethics: An Ecofeminist Perspective*. Lanham, MD: Rowman & Littlefield.

Kontogeorgopoulos, N. (1999) Sustainable tourism or sustainable development? Financial crisis, ecotourism, and the 'Amazing Thailand' campaign. *Journal of Sustainable Tourism* 2 (4), 316–332.

Kunoshima Co. Ltd. (2019) [Visitor statistics]. Unpublished raw data.

Ministry of the Environment (2016) Waga kuni no seitaikei nado ni higai o oyobosu osore no aru gairaishu risuto [A list of alien species potentially damage Japanese ecosystems]. https://www.env.go.jp/press/files/jp/26594.pdf (accessed March 2019).

McDonald, J.R. (2009) Complexity science: An alternative world view for understanding sustainable tourism development. *Journal of Sustainable Tourism* 17 (4), 455–471.

Moore, S.A. and Rodger, K. (2010) Wildlife tourism as a common pool resource issue: Enabling conditions for sustainability governance. *Journal of Sustainable Tourism* 18 (7), 831–844.

Morioka, T., Saito, O. and Yabar, H. (2006) The pathway to a sustainable industrial society: Initiative of the Research Institute for Sustainability Science (RISS) at Osaka University. *Sustainability Science* 1 (1), 65–82.

Moscardo, G. and Murphy, L. (2014) There is no such thing as sustainable tourism: Reconceptualizing tourism as a tool for sustainability. *Sustainability* 6 (5), 2538–2561.

Nassauer, J.I. and Corry, R.C. (2004) Using normative scenarios in landscape ecology. *Landscape Ecology* 19 (4), 343–356.

Newsome, D., Dowling, R. and Moore, S. (2005) *Wildlife Tourism*. Clevedon: Channel View Publications.

Notzke, C. (2016) Wild horse-based tourism as wildlife tourism: The wild horse as the other. *Current Issues in Tourism* 19 (12), 1235–1259.

Plumwood, V. (1993) *Feminism and the Mastery of Nature*. London: Routledge.

Rodger, K., Moore, S.A. and Newsome, D. (2007) Wildlife tours in Australia: Characteristics, the place of science and sustainable futures. *Journal of Sustainable Tourism* 15 (2), 160–179.

Russon, A.E. and Wallis, J. (2014) *Primate Tourism: A Tool for Conservation?* Cambridge: Cambridge University Press.

Saarinen, J. (2006) Traditions of sustainability in tourism studies. *Annals of Tourism Research* 33 (4), 1121–1140

Sakurai, F. (2016) The history of European rabbits which were brought to Japan (according to old Japanese paintings). *The Japanese Society of Veterinary History* 53, 24–31. http://jsvh.umin.jp/archives/pdf/53/053024031.pdf (accessed March 2019).

Sato, R. (2010) *Sea-Dumped Chemical Weapons: Japan.* Washington, DC: Global Green USA. See https://static1.squarespace.com/static/5548ed90e4b0b0a763d0e704/t/55548e55e4b0cef71eee4bd8/1431604821083/publication-114-1.pdf (accessed March 2019).

Schneider, F. and Rist, S. (2014) Envisioning sustainable water futures in a transdisciplinary learning process: Combining normative, explorative, and participatory scenario approaches. *Sustainability Science* 9 (4), 463–481.

Setouchi, M. (2017) *Nippon Nekojima Kikou* [*Japan Cat Island Travel*]. Tokyo: East Press.

Shani, A. and Pizam, A. (2008) Towards an ethical framework for animal-based attractions. *International Journal of Contemporary Hospitality Management* 20 (6), 679–693.

Smith, H., Samuels, A. and Bradley, S. (2008) Reducing risky interactions between tourist and free-ranging dolphins (*Tursiops* sp.) in an artificial feeding program at Monkey Mia, Western Australia. *Tourism Management* 29 (5), 994–1001.

Takehara City (2017) Heisei 29 nendo Takeharashi Tōkeisho [Takehara City's statistical documentation of 2017]. See http://www.city.takehara.lg.jp/soumu/toukei/toukeisyo.html (accessed March 2019)

Takeo, S. (2010) Hiroshima heiwa no tabi [Peace journey in Hiroshima]. *PRIME* 32, 131–135.

Tuan, Y.F. (1988) *Dominance and Affection: The Making of Pets* (S. Kataoka and H. Kaneri, trans.). Tokyo: Kōsakusha.

UNWTO (n.d.) Definition. See https://www.unwto.org/sustainable-development (accessed March 2019).

Usui, R. and Funck, C. (2017) Not quite wild, but not domesticated either: Contradicting management decisions on free-ranging sika deer (*Cervus nippon*) at two tourism sites in Japan. In I. Borges de Lima and R.J. Green (eds) *Wildlife Tourism, Environmental Learning and Ethical Encounters: Geoheritage, Geoparks and Geotourism* (pp. 247–261). Cham: Springer.

Usui, R. and Funck, C. (2018) The Role that Repeat Visitors Play in Managing Tourist–Rabbit Interactions for Wildlife Tourism on Ōkunoshima Island in Hiroshima, Japan. Paper presented at the 34th International Geographical Congress, Quebec City, Canada, 9 August.

Usui, R., Wei, X. and Funck, C. (2017) The power of social media in regional tourism development: A case study from Ōkunoshima Island in Hiroshima, Japan. *Current Issues in Tourism* 21 (18), 2052–2056.

van Notten, P.W.F., Rotmans, J., van Asselt, M.B.A. and Rothman, D.S. (2003) An updated scenario typology. *Futures* 35, 423–443.

Vance, L. (1997) Ecofeminism and wilderness. *NWSA Journal* 9 (3), 60–76.

Warren, K.J. (1990) The power and the promise of ecological feminism. *Environmental Ethics* 12 (2), 125–146.

Weaver, D. (2002) Asian ecotourism: Patterns and themes. *Tourism Geographies* 4 (2), 153–172.

Yeoman, I., Robertson, M. and Smith, K. (2011) A futurist's view on the future of events. In S. Page and J. Connell (eds) *The Routledge Handbook of Events* (pp. 507–525). Abingdon: Routledge. See https://www.tomorrowstourist.com/pdf/thefuturistsviewonthefutureofevents.pdf (March 2019).

Yeoman, I., Andrade, A., Leguma, E., Wolf, N., Ezra, P., Tan, R. and McMahon-Beattie, U. (2015) 2050: New Zealand's sustainable future. *Journal of Tourism Futures* 1 (2), 117–130.

Yudina, O. and Fennell, D. (2013) Ecofeminism in the tourism context: A discussion of the use of other-than-human animals as food in tourism. *Tourism Recreation Research* 38 (1), 55–69.

Zolfani, S.H., Sedaghat, M., Makmoon, R. and Zavadskas, E.K. (2015) Sustainable tourism: A comprehensive literature review on frameworks and applications. *Economic Research* 28 (10), 1–30.

4 Representing Wild Animals to Humans: The Ethical Future of Wildlife Tourism

Georgette Leah Burns and
Judith Benz-Schwarzburg

'As co-habitants on the same planet, humans and other animal species engage with each other constantly, and in a large variety of ways' (Burns & Paterson, 2014: ix). Wildlife tourism is one of the many spaces in which this human–animal relationship takes place. However, it is besieged by challenges. Given the current trajectory of wildlife population depletion (WWF, 2018), and the increasing pressure of global overtourism (Herntrei & Steckenbauer, 2018), the future of wildlife tourism initially looks grim. Between 1970 and 2014, species population sizes declined 60% (WWF, 2018), with high-profile species such as the blue macaw (Butchart *et al.*, 2018) and the northern white rhino (Gross, 2018) becoming extinct in the wild in 2018. In addition, human–wildlife interactions often take place in ways that are ethically problematic, evidenced in the high-profile attention given to this in a recent *National Geographic* report (Daly, 2019) and in travel agencies reacting to criticism from animal welfare and animal rights organisations (World Animal Protection, 2017). It is time to question: if wildlife tourism still exists in the future, what ethical challenges will it face?

Determining what is ethically appropriate in wildlife tourism is not simple and, until recently, has received scant attention in the scholarly tourism literature (Burns, 2017). The Wildlife Tourism Model proposed by Duffus and Deardon (1990), based on consideration of the ecology of the animal, the user/tourist and the historical context of their relationship, is missing a decisive animal ethics perspective. As such, it sheds no light on the emotional or psychological state of the animal and is limited as a tool to ethically evaluate wildlife tourism. Other models also seem limited. For example, animals are not considered in the 10 principles which comprise the World Tourism Organisation's (WTO) Global Code of Ethics for Tourism (Burns, 2015; Fennell, 2015). A rare exception is the seven principles approach devised by Burns *et al.* (2011) as an ecocentric framework for managing wildlife tourism. Searching for a way forward, we will integrate ideas from these ecocentric principles and the more general animal ethics debate, especially the zoo ethics debate, into the wildlife tourism debate to describe problems relevant to current trends in wildlife encounters. While we proceed, we will narrow our focus to some selected issues and aspects that seem paramount.

Firstly, our focus is solely on wildlife tourism in captive settings – where encounters are more likely to withstand the pressure of at least some of the aforementioned challenges. Secondly, we presuppose that the majority of human–wildlife encounters already are, and increasingly will be, designed and managed by humans: we rarely find non-orchestrated forms of wildlife encounters nowadays. Therefore, we aim to evaluate the forms and contents of representations bringing the wild animal to humans in settings such as zoos and shows. Thirdly, we exclusively refer to two ethical arguments that seem to dominate the debate about wildlife encounters: the argument that a carefully designed and executed wildlife encounter is of value because it serves species conservation and education with respect to conservation, and the argument that such encounters are ethically permissible as long as animal welfare is guaranteed. Although these two arguments seem widely accepted, this does not mean that they are the only valid and relevant ones in the debate and others are discussed elsewhere (Benz-Schwarzburg & Burns, under preparation).

These three boundaries of our discussion enable us to explore a range of ways that humans and animals currently interact. For that, we will repeatedly take into consideration case examples, including possible future wildlife tourism scenarios. By examining them, we highlight the importance of conservation and animal welfare as ethical norms, and propose not only what a more ethically responsible future could, and should, look like, but also what would be needed to achieve that goal.

Human–Wildlife Encounters in Captive Settings: Perspectives from Conservation and Welfare Ethics

When asking the question whether and to what extent wildlife tourism is ethically defensible, we might argue that encountering live animals can increase awareness of species and their conservation status. This conservation argument is at the heart of the main justification narrative of the World Association of Zoos and Aquariums and a key element in their strategy for the future (WAZA, 2019a). This is highly relevant as zoos are main destinations for tourists who seek to encounter wild animals. The argument has been challenged in the ethical debate. Zoos rarely provide in-depth information about conservation; their contribution to breeding and subsequent reintroduction to the wild seems limited; more convincing alternatives to zoos exist; and zoos fail to present animals with natural behaviour patterns in the first place (see e.g. Benz-Schwarzburg, 2020; Lee, 2005; Malamud, 2017). However, we will accept the argument in principle: if high-quality conservation and education on conservation takes place, it provides some ethical leverage in favour of zoos as places of wildlife encounters. Still, as entire species are the ethical target of this ecocentric argument, conservation ethics risks conflict with other ethical norms, such as biocentrism, that focus on the individual animal. While polar bears in zoos might be used as ambassadors for species that suffer from climate change, individual polar bears used as such are among the large, carnivore species whose welfare is severely compromised in captive settings (Clubb & Mason, 2003). This example highlights a possible clash of ethical norms, or possible pitfalls if we only concentrate on the conservation argument. Subsequently, scholars have turned to other norms, including those that highlight the perspectives and interests of animals as individual subjects.

One such norm, which has dominated the zoo debate to a considerable degree (WAZA, 2019b), is usually acknowledged by those defending zoos as much as by those criticising them: animal welfare. Animals in captivity clearly have less choice and control over their lives: all aspects of their lives are managed, allowed for or denied by human caretakers. This applies to night/day rhythms, feeding regimes, space provision, mating and breeding opportunities, and group composition. Furthermore, animals in zoos and other wildlife tourism contexts such as elephant rides, shows or petting programmes are subjected to a high degree of taming and training that is necessary to perform for tourists and to facilitate medical supervision. The high dependency of the animals comes with a great responsibility for their welfare on the side of caretakers, which is not always realised.

Animal welfare ethics argues against the background assumption that animals are morally relevant in themselves – meaning that they are of intrinsic value (e.g. Burns *et al.*, 2011) and do not matter only because they provide instrumental value for human ends. This is coupled with another assumption: animals matter because they are sentient beings. For these reasons, the so-called principle of non-maleficence asks us to avoid extensive unnecessary harm to animals (DeGrazia, 2005). The so-called Five Freedoms, originally developed for farm animals in the 1970s but often cited in wider contexts, state that an animal should be free from (1) thirst, hunger and malnutrition; (2) discomfort and exposure; (3) pain, injury and disease; (4) fear and distress; and (5) have the freedom to express normal behaviour (Farm Animal Welfare Council, 2012). They are clearly relevant to the presentation and actual treatment of animals in wildlife tourism where we are predominantly dealing with harm in terms of animal welfare. However, such consideration of animal welfare needs updating beyond the minimum standards in the Five Freedoms to include, for example, not just current knowledge on animal biology but also knowledge about what causes fear and distress (Mellor, 2016). Without these considerations, there is a high risk of underestimating the welfare harm done to many animals under human care, especially if they have complex emotional, social and cognitive needs (Benz-Schwarzburg, 2020). Thus, mindfulness of the precautionary principle, a popular idea in environmental ethics but largely absent from wildlife tourism considerations, is useful as it advocates that if consequences of an action are unknown then the action should not be undertaken (Burns *et al.*, 2011).

Both ethical arguments, conservation and welfare, provide us with useful moral values from which we can assess the future of wildlife tourism, and we will return to and make use of them throughout the chapter. Note that both values can be relevant directly or indirectly. We can ask whether husbandry, training or performance conditions in wildlife encounters directly violate them when it comes to animal treatment. However, their indirect relevance is often overlooked, which is why we want to emphasise it here. Conservation and welfare can be addressed in a problematic way or overridden by other issues also in abstract representations of animals, meaning in stories being told or pictures being provided to tourists – maybe even without a real animal being present and affected by the represented treatment at that very moment. Particularly, if entertainment is at the forefront, the norms of conservation and welfare can easily be overridden in the sense that visitors then care about them less (Schroepfer *et al.*, 2011).

We are ethically obliged not only to ask ourselves how animals are treated but also what the nature and content of the representation of the animal does to the values of conservation and welfare. The latter question turns our focus to the diversity of levels of ethical responsibility we face when it comes to wildlife tourism: there is the level of the species, highlighted by the conservation argument, and the level of the individual animal, addressed by welfare concerns. But there is also the level of the human–animal relationship. Wildlife tourism can do damage here as well. All levels of responsibility ultimately fall on those managing the encounters. The reflective manager principle enunciated by Burns *et al.* (2011) is useful here as it provides a guide for managers to reflect on their own ethical stances and the influence this has on their management styles.

Managing wildlife tourisms in a way that juggles the three objectives of conservation, welfare and entertainment is very challenging. Indeed, we seem to face a trend towards entertainment that threatens conservation and welfare (e.g. Beardsworth & Bryman, 2001; Benz-Schwarzburg, 2020; Carr & Cohen, 2011). In the next section, we will explain and discuss this trend to derive some more specific conclusions for when wildlife tourism turns into an ethically problematic encounter. We do so to prepare for our final discussion where we will argue that the main challenge for the future of wildlife tourism is creating an experience where entertainment is not a value in itself but a side effect of a carefully designed encounter that puts the animals, their conservation and welfare first. We question whether this scenario is attainable, as current trends seem to run counter to it. Nevertheless, we should, from an ethical perspective, insist on proposing a future where dual outcomes of human entertainment and wildlife conservation can be achieved.

Current Trends in Wildlife Tourism: When Disneyisation Overruns Ethics

Features of Disneyisation

Captive settings often create immersive experiences that are fake theme worlds (Beardsworth & Bryman, 2001: 87), providing high entertainment value for tourists for whom encountering wildlife is the main attraction of a fun-day-out package. This trend has been captured under the term of Disneyisation, whose features include theming, de-differentiation of consumption, massive merchandising and emotional labour performed by human and animal staff (Beardsworth & Bryman, 2001; Benz-Schwarzburg & Leitsberger, 2015).

Theming occurs as zoos seek to reinvent themselves as conservation parks that display animals in contexts 'more consistent with modern sensibilities and attitudes to animals in captivity' (Beardsworth & Bryman, 2001: 93). Thus, in many zoos, the theming of exhibits, as 'quasifications of natural habitats' (Beardsworth & Bryman, 2001: 92), is connected with messages about conservation. The newly opened (July 2019) Aquarium at Whipsnade Zoo, run by the Zoological Society of London, is an example of this as it is promoted as 'the UK's first aquarium dedicated to conserving the world's most astonishing and endangered freshwater fish' (Whipsnade Zoo, 2019). Although not without ethical concerns, particularly in terms of potentially fake representations that accordingly lead to fake expectations of wildlife, theming can have positive conservation and educational outcomes. The other features of Disneyisation, however, are often more ethically problematic.

As consumption is dedifferentiated, different forms become interlocked and difficult to distinguish (Bryman, 1999). For example, zoos are no longer places to just purchase a ticket for entry but increasingly contain extensive eating and shopping facilities. This growing commercialisation is tied with the third feature, of merchandising. Most zoos have an extensive range of merchandise which, like the other features, has expanded in importance since identified by Beardsworth and Bryman (2001). While often tied to messages about conservation – visitors to zoos can purchase images of iconic and threatened species or products supposedly to assist them in the wild – this kind of rendering animals into merchandising products again clearly increases the commodification of the animals and the commercialisation of the entire encounter.

Emotional labour, the final feature of Disneyisation, refers to the expectation of zoo employees to provide service with a smile. Although originally conceived for people, 'zoos frequently enlarge the field of emotional labour by conscripting their animal inmates ... into the performance' (Beardsworth & Bryman, 2001: 97). In this representation, animals are trained to display behaviour that can be interpreted as expressing a particular emotion; for example, when a sea lion kisses a keeper the audience sees affection which in turn affects human emotion about the experience.

Beardsworth and Bryman (2001) wrote of evidence of potential Disneyisation in zoos and we can see, in the intervening years, the four features becoming entrenched in wildlife tourism in general. Disneyisation illustrates an increasing turn in wildlife tourism towards entertainment and consumption, in which animals are objectified as attractions complementing an entertainment package composed for the visitor as a consumer. As such, it is rarely compatible with the ecocentric principles posed by Burns et al. (2011), or in general with the claim to put the animal first. Theming, dedifferentiation of consumption and merchandising lead to a perception of the animal as a consumable product that exists to entertain us, in turn providing a legitimating framework for an exploitative attitude. The requirement of emotional labour from the animal, often a crucial component of captive animal performances, is a key example of such exploitation for entertainment. This trend is problematic from a conservation and welfare perspective as it turns the focus away from the welfare perspective of the animal and the conservation perspective of the species and towards the entertainment interests of the visitors. Using some case examples, we discuss this in more detail in the next section.

Current examples of Disneyisation

Among the tourism attractions exclusively designed for entertainment purposes are shows where animals are trained to perform for humans. At the Samui Monkey Theatre in Ko Samui, Thailand, for example, chained macaques play musical instruments in staged performances marketed as ones that will 'make you laugh': 'you will have a good time here' (Discovery Thailand, 2019). The 'good time' is devoid of any conservation messages, presenting these animals as existing to perform for human enjoyment. Similarly, at the Phuket Zoo in Thailand, elephants entertain visitors by playing football, dancing and painting (Phuket Directory, 2020): all unnatural activities that usually involve the inhumane breaking-in of elephants (Cohen, 2015). In Thailand, the relationship between humans and elephants has a strong anthropocentric history (Servaes, 2017) of elephants being used for commercial gain in the logging industry. This is also true for the macaques, trained to harvest

Figure 4.1 Elephant performing at Phuket Safari Eco+ in Thailand (Photographer: Aaron Gekoski)

coconuts. Their representation in the tourism industry, as continuing unpaid workers, is an extension of this power relationship (Figure 4.1).

Many zoos offer shows as well; for example, presenting animals feeding while a keeper provides information about them. While welfare issues concerning husbandry and training might be of lesser concern here, these shows still add to the described problematic trends. They are, in fact, rarely purely educational and rarely devoid of fun-for-people elements. For example, at Sea World in Australia, animals are trained to perform tricks and we can still see people riding dolphins and seals kissing keepers (Sea World, 2019).

These examples demonstrate that the trend of Disneyisation can be ethically scrutinised not only from a welfare perspective but also as indirectly affecting animal welfare. How the animal actually feels becomes irrelevant to the visitor. The animals that become staff members are trained to promote positive emotions in the visitors by displaying fake emotions themselves. The ever-smiling, human entertainer is thus accompanied by an animal entertainer who, regardless of its own emotional state, has to make the visitor feel good and have fun (Benz-Schwarzburg & Leitsberger, 2015: 27–29). Furthermore, the behaviour shown is very unnatural: it orientates on human aesthetic preferences, not only behaviour-wise but often also costume-wise, and encourages the human consumer to ridicule the struggling but clumsy animal performer – as seen at the Samui Monkey Theatre. In many cases, training methods probably included punishment and negative reinforcement. But even if they did not, the visitors' perception of the animals as subjects with individual welfare that matters is distorted by the animal performer who, in the visitor's perception, is happily cooperating.

The Future of Wildlife Tourism

While it is encouraging that welfare criticism has reached a point where some travel companies no longer offer tours to shows like those described above (World Animal Protection, 2017), they still exist. For zoos, they also verify the future as predicted by Beardsworth and Bryman (2001), a future that increasingly mixes the serious ideals of zoos – such as conservation and education – with entertainment and artificiality.

Beardsworth and Bryman (2001) assumed that, given this trend, visitors will encounter increasingly more technologically created animals instead of real ones in such hybrid zoos. This trend smoothly fits with today's plethora of semi-artificial, already technologically supported, experiences. Examples of this development can be found around the globe. The Ghongzou Zoo in China, for example, opened 'the world's one and only VR Zoo' in 2018: an exhibition encompassing encounters with virtual animals from the panda to the dinosaur (New China TV, 2018). In Basel, Switzerland, Vision Nemo was initiated by a Swiss foundation to propose an alternative to a new aquarium by creating an exhibit without living animals that brings the oceans and their inhabitants to the visitors by means of gigantic three-dimensional screens (Borgards, 2019). In general, zoos increasingly seem to work with smaller, embedded virtual tours and other means. For example, the Tiger Trek at Taronga Zoo in Australia is an immersive exhibit that takes visitors on a simulated flight to Sumatra where they disembark to view tigers in an enclosure that appears very natural before being prompted on exit to make shopping choices based on sustainable palm oil consumption (Kelly, 2018). Enclosures such as these follow the imperative behind the belonging in nature principle by encouraging visitor awareness of the natural habitat of the wildlife (Burns *et al.*, 2011). As such they can promote conservation.

Additionally, zoos have done much with regard to improving animal welfare. Interestingly, this also sometimes means getting rid of animals. At least some zoos no longer keep species which they cannot sustain in an appropriate way, or change to non-contact exhibits, also for ethical reasons. In Germany, Frankfurt Zoo (2019), for example, abandoned elephant husbandry in the early 1980s, and Cologne Zoo (2019) switched to keeping elephants in large groups with reduced contact between animals and caretakers. Greater specialisation and fewer animals are key components when it comes to the ethical future of zoos.

The described trend of getting rid of live animals can also be observed in other areas of wildlife tourism outside of zoos and aquaria. In 2017,

Circus Roncalli launched the animal-free show Storyteller (Roncalli, 2019a) which includes holographic horses, goldfish and elephants created by 3D projections using laser beams and cloud computing. The show also displays other artificial animals (e.g. a robot dog) and humans wearing animal costumes. Changing the show accordingly cost more than €500,000 but it also attracted more than 600,000 visitors in one year (Roncalli, 2019b). Circus director Bernhard Paul argues that having live circus animals is no longer contemporary and he gives ethical reasons for this, such as welfare concerns with transporting animals over long distances in trucks increasingly stuck in traffic (Nachrichten.at/apa, 2018). Animal Fair lists nine other prestigious circuses, including Cirque du Soleil, that also work animal-free (Animal.fair, 2019).

Such future scenarios remove the real animal from display and thus from experiencing and possibly suffering harm through, for example, captive husbandry, hand-raising, training via punishment and abusive treatment by trainers and visitors. However, the indirect level of welfare relevance, which we emphasised in our section on ethical norms, could still be problematic: display of non-live, or unreal, animals for human entertainment is not new and bares some possible pitfalls. For centuries, we have been entertained by puppet shows depicting animal performers. For decades, we have watched images of animals in anthropomorphised cartoons and identified with them. Recent developments in virtual reality and augmented reality should be careful not to continue with ethically problematic contents of representations such as anthropomorphism and – maybe even more problematic – anthropocentrism. Such contents conflict with an idea of animals as sentient beings with intrinsic value whose individual welfare we should care about.

Final Discussion: Reflections on Representation, Relationship, Welfare and Conservation

Animal welfare and the content of representations

In highly orchestrated captive settings, tourists are confronted with a specific representation of the tamed or exotic wild by the content of the show which might be brought to them by living animals or, increasingly, by non-living components. As we have seen, this is true not only for zoos but also for circuses, theme parks and other tourism venues that offer animal shows as their defining product and focus even more directly on entertainment. Often, shows are only one part of a more comprehensive encounter-the-live-animal programme that also includes components such as feeding the animals, petting them or riding on them.

Where animals are reduced in status to an economic commodity, performing for humans to make money for the presenter, then treatment as non-sentient, lifeless or less-than human is common (Burns, 2015; Linzey, 2009). As an object of tourism, the animal being presented is not only available to the tourist gaze but also becomes a marketable item that can be bought and sold and, at least visually, consumed (Burns, 2015). This type of objectifying highlights an anthropocentric approach concerned only with the instrumental value of the animal for human entertainment (Burns, 2015). We think that this kind of instrumentalisation as consumable entertainment object is also apparent in representations of virtual animals. In the new Roncalli show, for example, the non-real animals are still displayed as performing for tourists, with the primary objective to entertain people and make a profit for the company. Despite not being real, the message conveyed is still that consumption of animals in this way, in a human–animal relationship that affords ultimate power and control to people, is acceptable.

Strengthening visitor perception of the animal's perspective, as something that matters, can be achieved by presenting and representing animals as cognitively, emotionally and socially complex beings. This presupposes letting animals perform natural behaviours and educating zoo visitors about ethological needs: needs that captivity can rarely fulfil. This approach requires reflection and self-criticism from the institutions, up to the point where entire institutions or specific exhibits and shows are phased out for ethical reasons.

For the entertainment industry, current trends – for example, the diminishing number of live animal circuses, and advances in technology – suggest that they might continue to shy away from traditional shows and instead rely on artificial animals such as robots, 3D projections and virtual reality settings. This solves some ethical problems but adds to the trend that builds entertainment by means of (mis)representing animals. Cases of (mis)representation can, for example, be characterised by diminishing or disregarding norms such as conservation and welfare that are often not a focus of the content of what the artificial animals represent.

Animal welfare and the human–animal relationship

Changes in our relationships with all animals, not just in the arena of wildlife tourism, towards recognition and consideration of their intrinsic value are clearly gaining momentum. Legal recognition of animals as sentient beings in New Zealand in 2015 is one example (New Zealand Parliament, 2015). So, too, is the passing of a Senate Public Bill in Canada in 2019 ending the captivity of whales and dolphins (Parliament of Canada, 2019), and the 2019 opening of an elephant sanctuary in Thailand where tourists are not permitted to interact directly with the elephants (World Animal Protection, 2019).

Such developments are direct reactions to ethical demands and should be taken seriously and implemented according to their intention. Elephant tourism in Thailand, for example the Maetaeng Elephant Park (2020), where the animals paint for and/or are ridden by tourists, has long been criticised on the basis of potential harm (Servaes, 2017). Non-contact encounters seem to reduce harm to the animals, for example because training elephants to tolerate human riders is no longer necessary. But reducing direct contact between animals and tourists is also an answer to a new perception of animals that are not there to entertain us, to be touched and petted, ridden and fed by us, in the first place. Elements supportive of a trend towards entertainment and Disneyisation should be reduced and avoided in wildlife tourism. Such activities clearly contradict the idea of the intrinsic value of animals as well as the kind of respect they deserve as sentient beings.

Zoo exhibits can also change along ethical lines, at least to some degree. The design of enclosures should accommodate species needs rather than visitors wants. In the Bronx Zoo (2019) in New York, visitors can take a tour on the Wild Asia Monorail which transports them over the exhibits. Such rides could, in the future, recognise visitors as potential intruders and force them to maintain distance from the wildlife, with no guarantee of seeing an animal. The Masoala Rainforest at Zurich Zoo (2019) in Switzerland attempts something similar. In these types of exhibits emphasis is on visitors observing the animals on the animals' terms – if they want it. Such displays demand respect, patience and humbleness from the visitor and ideally create this expectation through their marketing campaigns.

Conservation and the human–animal relationship

While shows attract entertainment-oriented visitors to encounter wildlife, they also distract from serious messages such as conservation (Schroepfer *et al.*, 2011). In addition, even if entertainment is pushed back and conservation is brought forward in animal representations, the choice of messages conveyed needs careful monitoring. For example, while the Tiger Trek at Taronga Zoo is an attractive and entertaining exhibit for visitors that simultaneously attempts to educate them about the connection between tiger conservation and their personal consumption of palm oil products, the solution to the problem offered, that is, certified palm oil, is a questionable one, not only because replacing the amount of palm oil we use with sustainably produced palm oil seems impossible, but also because the areas used today for certified palm oil were once intact rainforest. Thus, Borneo Orangutan Survival Foundation (BOSF), an experienced non-governmental organisation (NGO) in the field in Indonesia, promotes a ban on palm oil. Thus, the Tiger Trek example illustrates how zoos sometimes create exhibits that fuse exciting encounters with an exotic and rare species with superficial conservation messages. Conservation aims will only be reached if wildlife tourism substantially enhances the pedagogical quality of programmes and chooses from the outset an approach that puts ethical considerations at the core of the planning of such programmes.

Today, a 'good zoo' is recognised by its focus on education and conservation (Gray, 2017; Ward, 2016). Zoos could further enhance their positive ethical position by encompassing principles of moral obligation and reasoning in their education of visitors (Burns *et al.*, 2011). These principles enable increased engagement of visitors with conservation messages and understanding of animal welfare issues. Thus, the ethical turn we should pursue consists of an increasing incorporation of conservation messages, which implore the tourist to consider not just the animal as an object of entertainment but also the animal's own perspective as a living being: a being that not only belongs in, and stands in a relationship with its natural habitat, but that also often stands in an exploitative relationship with humans.

Recommendations for Future Captive Wildlife Tourism

In our preferred, more ethically sound, future for wildlife tourism, the trend towards Disneyisation, recognised as incompatible with an appropriate human–animal relationship, is halted and ideally reversed. Instead of focusing on entertainment for people, through theming, merchandising, dedifferentiation of consumption and trained emotional labour of animals, we need to step away from exploitative aspects and instead turn to conservation and welfare as the main values of captive wildlife tourism.

Live animals should only be involved where welfare concerns are solved. Note that captivity is, to some degree, always coupled with subordinating the animal's welfare interests to the interests of the visitor. Perhaps virtual representations of animals are the answer to combine human desire for relationships with wildlife with non-exploitative ethics. But this too needs awareness that the tourism sector co-shapes the interests and expectations of the visitors by the type and content of the representations they create, virtual or otherwise.

Our vision for the future is a setting where animals reside in habitats as close to nature as possible, and where viewing by tourists is unobtrusive on animal lives as well as informative to the visitor from a conservation perspective. Where this cannot be guaranteed, such as in the case of captive large marine mammals, displays are only virtual. In this future, live wildlife is encountered as something very special and valuable, and expensive: comparable to seeing a unique art exhibition where visitors apply for tickets in advance and are guided by experts instead of having unlimited and unguided access at their convenience. Animal encounters are not the place for Disneyised entertainment. This detracts from any serious conservation message and provides conflicting ideas about the human–animal relationship. Entertainment as a main objective in the future is left to entertainment parks, which do not contain live animals to ensure no welfare harm is done as well as to ensure that the ideas of entertainment and using sentient animals are not conflated. Shows, such as circuses, may still exist but without live animals – following the model used by Roncalli to embrace the possibilities of new technology with careful thinking about the content of such representations. The aim is a kind of representation that sparks a deep and resounding interest in the individual animal, its natural behaviour and the norms of animal welfare and conservation.

Conclusion

As the types and numbers of wildlife species decline, captive tourism settings become increasingly important as places to encounter what remains. However, we need to develop a substantially more critical ethical perspective on current wildlife tourism practices, and supposedly revolutionary technological ones, to ensure a future that embraces a human–animal relationship that is positive for both people and animals. Thus, we need to remain critical of the way we display and represent wildlife for human consumption, be cautious of commercialisation that distracts from animal-centred values and place consideration of animal conservation and welfare at the forefront of any wildlife tourism enterprises.

Acknowledgements

Part of this research was funded by the FWF (P 31466). We are very grateful to photographer Aaron Gekoski for giving us permission to use his elephant photograph. For more of Gekoski's wildlife tourism pictures see www.aarongekoski.com.

References

Animal.fair (2019) Zirkus – hier lacht nur der Clown. See www.animalfair.at/infothek/zirkus-hier-lacht-nur-der-clown (accessed September 2019).
Beardsworth, A. and Bryman, A.E. (2001) The wild animal in late modernity: The case of the Disneyization of zoos. *Tourist Studies* 1 (1), 83–104.
Benz-Schwarzburg, J. (2020) *Cognitive Kin, Moral Strangers? Linking Animal Cognition, Animal Ethics & Animal Welfare*. Leiden: Brill.
Benz-Schwarzburg, J. and Leitsberger, M. (2015) Zoos zwischen Artenschutz und Disneyworld. *Tierstudien* 7, 17–30.
Benz-Schwarzburg, J. and Burns, G.L. (in preparation) Ethics in wildlife tourism: Considering the future.
Borgards, R. (2019) Der virtuelle Zoo. Unterwegs zum zoologischen Datengarten. In I. Bolinski and S. Rieger (eds) *Das verdatete Tier. Zum Animal Turn in der Medienwissenschaft* (pp. 139–150). Berlin: J.B. Metzler.
Bronx Zoo (2019) Wild Asia monorail. See https://bronxzoo.com/rides/wild-asia-monorail-seasonal (accessed September 2019).
Bryman, A. (1999) The Disneyization of society. *The Sociological Review* 47, 25–47.
Burns, G.L. (2015) Animals as tourism objects: Ethically refocusing relationships between tourists and wildlife. In K. Markwell (ed.) *Animals and Tourism: Understanding Diverse Relationships* (pp. 44–59). Bristol: Channel View Publications.

Burns, G.L. (2017) Ethics and responsibility in wildlife tourism: Lessons from compassionate conservation in the Anthropocene. In R. Green and I. Lima (eds) *Wildlife Tourism, Environmental Learning and Ethical Encounters* (pp. 213–220). Cham: Springer.

Burns, G.L. and Paterson, M. (2014) Introduction. In G.L. Burns and M. Paterson (eds) *Engaging with Animals: Interpretations of a Shared Existence* (pp. ix–xiv). Sydney: Sydney University Press.

Burns, G.L., Macbeth, J. and Moore, S. (2011) Should dingoes die? Principles for engaging ecocentric ethics in wildlife tourism management. *Journal of Ecotourism* 10 (3), 179–196.

Butchart, S.H.M., Lowe, S., Martin, R.W., Symes, A., Westrip, J.R.S. and Wheatley, H. (2018) Which bird species have gone extinct? A novel classification approach. *Biological Conservation* 227, 9–18.

Carr, N. and Cohen, S. (2011) The public face of zoos: Images of entertainment, education and conservation. *Anthrozoös* 24 (2), 175–189.

Clubb, R., and Mason, G. (2003) Captivity effects on wide-ranging carnivores. *Nature* 425 (6957), 473–474.

Cohen, E. (2015) Young elephants in Thai tourism: A fatal attraction. In K. Markwell (ed.) *Animals and Tourism: Understanding Diverse Relationships* (pp. 163–177). Bristol: Channel View Publications.

Cologne Zoo (2019) Cologne Zoo elephant park. See www.koelnerzoo.de/en/tiere-2#elefantenpark (accessed June 2019).

Daly, N. (2019) Suffering unseen: The dark truth behind wildlife tourism. *National Geographic*, June 2019. See www.nationalgeographic.com/magazine/2019/06/global-wildlife-tourism-social-media-causes-animal-suffering/ (accessed November 2019).

DeGrazia, D. (2005) Regarding the last frontier of bigotry. *Logos* 4 (2). See http://www.logosjournal.com/issue_4.2/degrazia.htm (accessed 12 August 2020).

Discovery Thailand (2019) Samui Monkey Theater in Koh Samui. See www.discovery-thailand.com/Koh_Samui_Samui_Monkey_Theater.asp (accessed September 2019).

Duffus, D.A. and Deardon, P. (1990) Non-consumptive wildlife-oriented recreation: A conceptual model. *Biological Conservation* 53, 213–231.

Farm Animal Welfare Council (2012) Five Freedoms. See https://webarchive.nationalarchives.gov.uk/20121010012427/http://www.fawc.org.uk/freedoms.htm (accessed September 2019).

Fennell, D. (2015) The status of animal ethics research in tourism: A review of theory. In K. Markwell (ed.) *Animals and Tourism: Understanding Diverse Relationships* (pp. 27–43). Bristol: Channel View Publications.

Frankfurt Zoo (2019) Frankfurt Zoo. See https://www.zoo-frankfurt.de/faq (accessed 17 June 2019).

Gray, J. (2017) *Zoo Ethics: The Challenges of Compassionate Conservation*. London: CSIRO Publishing.

Gross, M. (2018) Last call to save the rhinos, *Current Biology* 28 (1), 1–3.

Herntrei, M. and Steckenbauer, G.C. (2018) Overtourism: A contribution towards the development of a conceptual model for retaining tourism acceptance within the tourism destinations. In D. Gursoy, S. Deesilatham and P. Piboonrungroj (eds) *Proceedings of the 8th Advances in Hospitality and Tourism Marketing and Management* (pp. 244–246). Bangkok: University of the Thai Chamber of Commerce.

Kelly, A. (2018) Inspiring pro-conservation behavior through innovations in zoo exhibit and campaign design. Master of Science thesis, Kansas State University.

Lee, K. (2005) *Zoos – A Philosophical Tour*. New York: Palgrave MacMillan.

Linzey, A. (2009) *Why Animal Suffering Matters: Philosophy, Theology, and Practical Ethics*. New York: Oxford University Press.

Malamud, R. (2017) The problem with zoos. In L. Kalof (ed.) *The Oxford Handbook of Animal Studies* (pp. 397–410). Oxford/New York: Oxford University Press.

Maetaeng Elephant Park (2020) Activities and shop. See https://www.maetaengelephant-park.com (accessed 12 August 2020).

Mellor, D.J. (2016) Updating animal welfare thinking: Moving beyond the 'five freedoms' towards 'a life worth living'. *Animals* 6, 21. doi: 10.3390/ani6030021.

Nachrichten.at/apa (2018) Circus Roncalli verzichtet jetzt auf Tiere und Plastik. See www.nachrichten.at/kultur/Circus-Roncalli-verzichtet-jetzt-auf-Tiere-und-Plastik;art16,3004004 (accessed September 2019).

New China TV (2018) LIVE: Welcome to world's one and only VR Zoo in S China's Guangzhou. Broadcast live on 4 January. See www.youtube.com/watch?v=71qTcaIRDB8 (accessed September 2019).

New Zealand Parliament (2015) Animal Welfare Amendment Bill. See https://www.parliament.nz/en/pb/bills-and-laws/bills-proposed-laws/document/00DBHOH_BILL12118_1/animal-welfare-amendment-bill (accessed September 2019).

Parliament of Canada (2019) Senate Public Bill. See https://www.parl.ca/LegisInfo/BillDetails.aspx?Language=e&Mode=1&billId=8063284&View=5 (accessed September 2019).

Phuket Directory (2020) Businesses and services: Phuket Zoo. See https://phuketdir.com/phuketzoo (accessed 12 August 2020).

Roncalli (2019a) Roncalli Circus. See www.roncalli.de (accessed September 2019).

Roncalli (2019b) #holographic #circustheaterroncalli #holograohicanimals #tierfrei #vegan #auchvegan #moderntimes @CircusRoncalli. Twitter Announcement from 19 April. See https://twitter.com/circusroncalli/status/1119229107265572864 (accessed September 2019).

Schroepfer, K.K., Rosati A.G., Chartrand, T. and Hare, B. (2011) Use of 'entertainment' chimpanzees in commercials distorts public perception regarding their conservation status. *PLoS ONE* 6 (10), e26048. doi:10.1371/journal.pone.0026048.

Sea World (2019) Seal guardians presentation. See https://seaworld.com.au/attractions/shows-and-presentations/seal-guardians-presentation (accessed September 2019).

Servaes, L. (2017) Elephants in tourism. Sustainable and practical approaches to captive elephant welfare and conservation in Thailand. In P. Malikhao (ed.) *Culture and Communication in Thailand* (pp. 127–138). Singapore: Springer.

Ward, S. (2016) How captivity helps conservation. *The Conversation*. See https://theconversation.com/in-defence-of-zoos-how-captivity-helps-conservation-56719 (accessed October 2019).

WAZA (2019a) Committing to conservation. The world zoo and aquarium conservation strategy. See www.waza.org/wp-content/uploads/2019/03/WAZA-Conservation-Strategy-2015_Landscape.pdf (accessed October 2019).

WAZA (2019b) WAZA code of ethics and animal welfare. See www.waza.org/wp-content/uploads/2019/05/WAZA-Code-of-Ethics.pdf (accessed August 2019).

Whipsnade Zoo (2019) News: An underwater world of wonder. See https://www.zsl.org/zsl-whipsnade-zoo/news/an-underwater-world-of-wonder-0 (accessed October 2019).

World Animal Protection (2017) Over 160 travel companies commit to end sale of elephant rides and shows. See https://www.worldanimalprotection.us/news/over-100-travel-companies-commit-end-sale-of-elephant-rides-and-shows (accessed September 2019).

World Animal Protection (2019) Thai elephant venue reopens without cruelty. See https://www.worldanimalprotection.org.au/news/thai-elephant-venue-reopens-without-cruelty (accessed 15 August 2020).

WWF (2018) *Living Planet Report – 2018: Aiming Higher.* M. Grooten and R.E.A. Almond (eds). Gland: World Wildlife Fund.

Zurich Zoo (2019) Masoala Rainforest. See www.zoo.ch/en/plan-your-visit/exhibits/masoala-rainforest (accessed September 2019).

Part 2
Human–Animal Encounters

5 The Rise of Selfie Safaris and the Future(s) of Wildlife Tourism

Jessica Bell Rizzolo

Introduction

Global participation in wildlife tourism is widespread, with tourism one of the primary mechanisms through which people interact with wildlife. The international market for wildlife tourism is valued at US$45 billion, with an annual growth rate of 10%, and it is expected to proliferate further with increases in global education and income (Newsome & Rodger, 2013). However, 'wildlife tourism' is a broad term that encompasses activities varied in their environmental and social impacts. Wildlife tourism includes diverse activities such as wildlife viewing; using wildlife for entertainment, transportation, trekking, hunting and fishing; visiting protected areas and sanctuaries; and visiting zoos and aquariums (Newsome *et al.*, 2005). Wildlife tourism is often classified along three dimensions. First, wildlife tourism can occur in captive, semi-captive or wild environments (Tisdell & Wilson, 2012). Second, wildlife tourism ventures differ in the relative emphasis they place on conservation or entertainment (Fennell, 2012; Shackley, 1996). Third, wildlife tourism is often categorised as consumptive or non-consumptive, with non-consumptive wildlife tourism defined as human recreation that does not remove or permanently alter wildlife (Duffus & Dearden, 1990).

The perception of wildlife tourism photography as benign for wildlife has contributed to the rise of selfie safaris. Selfies are photographs in which the self is the primary focus; these photographs are designed to be shared and to showcase the self in relation to some secondary product, such as a landscape or an animal (Lim, 2016). This chapter defines a 'selfie safari' as a form of wildlife tourism in which visitors have direct physical contact with wildlife in order to document and share this experience through a selfie. This form of wildlife tourism is contributing to extensive wildlife poaching, trafficking and mortality (Daly, 2017a; World Animal Protection [WAP], 2017). Documentation of these severe impacts contradicts the traditional conceptualisation of live animal encounters as non-consumptive, as it can be argued that tourist activities such as posing for selfies with tigers has permanent impacts on wildlife. However, given the nascent state of research on selfie safaris, uncertainty remains about the future of this form of wildlife tourism. Scenario planning allows tourism researchers to examine issues characterised by uncertainty and complexity, and to develop options for more sustainable futures (Duinker & Greig, 2007; Page et al., 2010).

I begin by discussing the methodology of scenario planning used in this chapter and contextualising it within the futures literature. Following Schoemaker's (1993) established method for scenario development, I examine: (a) the issue of selfie safaris, (b) the major actors who have an interest in this issue, (c) trends that will affect this issue and (d) key uncertainties whose resolution will affect the issue. Then, again building on Schoemaker (1993), I construct a utopian scenario, a dystopian scenario and other plausible scenarios for the future(s) of selfie safaris. Finally, I examine the ethical, management and social implications of these scenarios. One of the implications of scenario planning is that numerous futures are possible (Yeoman & Postma, 2014). Thus, I consider how these scenarios can inform a new model of non-consumptive wildlife tourism that builds a more sustainable future for wildlife.

Future Scenarios in Tourism

This chapter uses scenario planning to examine how selfie safaris may evolve over time, and to consider their potential impacts on wildlife and tourism. Future studies are based on the notion that there are multiple potential futures. This allows for the delineation and the generation of alternative and preferred futures (Robertson & Yeoman, 2014). Scenario planning is a systematic method for mapping these various possibilities and the decision points that influence them. It is especially useful for complex topics characterised by uncertainty (Robertson & Yeoman, 2014). This method incorporates information on the past and present with variables that will influence the topic of interest over time to develop potential scenarios. Scenarios are related to the concepts of signposts and signals. Signposts are evidence of scenarios occurring now, whereas signals are indicators of what the future could be (Yeoman & Postma, 2014). While there are multiple forms of scenario development, Schoemaker (1993) provides one of the most comprehensive and frequently cited methods (Amer et al., 2013). The scenarios in this chapter were developed through steps derived from Schoemaker's (1993) method (Figure 5.1), beginning with a definition of the issue of selfie safaris.

Selfie Safaris

The growth of wildlife selfies has been exponential. The number of wildlife selfies on Instagram (one of the primary social media platforms for selfies) rose 292% from 2014 to 2017, and over 40% of these selfies

1. Define the issue you wish to understand better.

2. Identify the major actors who would have an interest in this issue and describe their current roles.

3. Make a list of current trends that will affect the issue; for each, explain how and why it exerts an influence.

4. Identify key uncertainties whose resolution will significantly affect the issue.

5. Construct two forced scenarios by placing all positive outcomes of key uncertainties in one scenario, and all negative outcomes in another.

6. Assess the internal consistency and plausibility of these scenarios.

7. Eliminate scenarios that are not credible or impossible.

8. Using different outcomes of the key uncertainties, construct at least two more scenarios that bracket a wide range of outcomes.

9. Repeat steps 6 and 7 for the other scenarios.

10. Identify how major actors would behave in each of the scenarios.

Figure 5.1 Steps for scenario development (Derived from Schoemaker, 1993)

displayed a problematic level of proximity between the person and the wild animal (WAP, 2017). The locations of these selfies spanned continents and species, with most posts emerging from Australia, the United States, the United Kingdom, Thailand, Indonesia, Canada, India and Brazil. Animals used for wildlife selfies include elephants, sloths, koalas, tigers, lions, turtles, macaque monkeys, gibbons, orangutans, slow lorises, dolphins, snakes, anteaters, ocelots, toucans and manatees (WAP, 2017). This is a signpost that selfie safaris constitute a growing segment of global wildlife tourism and that multiple species are impacted. This signpost gives off various signals about how this form of wildlife tourism will affect wildlife and tourism over time. Those signals include negative animal welfare effects, the inability of tourists to accurately judge the repercussions of their actions and detrimental conservation impacts.

Wildlife is kept in captive conditions (Figure 5.2) or offered food as encouragement to come close to tourists (Figure 5.3). The latter, known as baiting, is associated with negative effects on wildlife such as habituation to humans and increased aggressiveness towards conspecifics, as well as problematic behaviour by tourists such as striking the animals or feeding them inappropriate food (D'Cruze et al., 2017). Both captivity-based

Figure 5.2 A tourist poses for a photo with a tiger in Thailand. At some facilities the adults are kept on a leash to be controlled or confined to small cages when not in use. It is suspected, but not proven, that drugs are used to keep the tigers docile in some attractions. World Animal Protection believes that animals should stay in the wild and not be used for entertainment (© World Animal Protection)

Figure 5.3 Tourists can pay for harmful close encounters and selfies with the Amazon river dolphins (*boto*) in Manaus, Brazil (© World Animal Protection/Nando Machado)

and baiting-based venues allow tourists a problematic level of physical proximity to wildlife.

Although the tourists who participate in selfie safaris may not intend to harm wildlife, a comparison of tourist perceptions with objective ratings of wildlife tourism attractions found that tourists are poor judges of the animal welfare and environmental impacts of their choices (Moorhouse et al., 2015). A review of the animal welfare conditions of wildlife attractions in Thailand (a primary destination for selfie safaris) found that 71% of these venues provided no conservation education to visitors and that severely inadequate welfare was common, with 75% of macaques, 86% of elephants and 99% of tigers held in deficient conditions (Schmidt-Burbach et al., 2015). Animals used for these venues commonly display symptoms of stress due to excessive handling, abuse, untreated wounds, lack of enrichment and disruption of natural behaviours (Schmidt-Burbach et al., 2015; WAP, 2017). Species easily harmed by frequent handling, such as sloths, are particularly vulnerable to the negative repercussions of selfie safaris (Carder et al., 2018); one report estimated that sloths taken from the wild for selfies may not survive longer than six months (WAP, 2017).

In addition to their adverse animal welfare impacts, selfie safaris pose a number of risks to the conservation of endangered species. Many of the species featured in selfies are endangered or threatened (WAP, 2017). Using these animals for selfie safaris is associated with wildlife trafficking, increased mortality (for both the target animal and, if captured from the wild, members of that animal's family) and habituation to humans (which makes rehabilitation and release difficult). Further, the creation and propagation of these selfies can impact human attitudes in a manner detrimental to conservation. These selfies normalise wild animals as 'cute' pets and photo props, which can help fuel the exotic pet trade (Kitson & Nekaris, 2017; Mutalib, 2018). For example, one of the most common reactions to a video of a slow loris being 'tickled' in captivity was 'I want one!' (Nekaris et al., 2013). The impacts of selfies may also extend to other forms of wildlife consumption. Even after controlling for age, gender, nationality and cognitions towards wildlife, posing for a wildlife selfie doubles the odds of eating/drinking wildlife and of buying wildlife products such as souvenirs made from wild animal parts (Rizzolo, in preparation). Overall, selfie safaris have substantial impacts on both animal welfare and conservation, and it is important for those interested in sustainable tourism to investigate different futures for wildlife selfies.

Actors

The three core actors who influence wildlife tourism are the tourist industry, tourists and governments. The tourist industry includes both those who own and operate wildlife tourism attractions as well as sites such as Tripadvisor that facilitate wildlife tourism. It is common for wildlife tourism providers to greenwash: to falsely claim that the attraction benefits conservation (Moorhouse et al., 2017b; Rizzolo, 2017). In many tourist locales, wildlife selfies are widespread and actively encouraged. Captive wildlife attractions are prevalent in Thailand (Schmidt-Burbach et al., 2015) and 54% of Latin American wildlife attractions offer direct contact with wild animals; in some locations, such as Manaus, Brazil, the number of tourism operators offering close contact with wild animals reaches 94% (WAP, 2017).

Tourists have a difficult time ascertaining the actual conservation impacts of the wildlife attraction and their decision to attend a venue is heavily influenced by travel sites such as Tripadvisor (Moorhouse et al., 2015, 2017b). However, Tripadvisor reviews are not a reliable indicator of animal welfare or conservation benefits. Most wildlife attractions with objectively poor standards for animal welfare and conservation still have primarily good reviews on Tripadvisor (Moorhouse et al., 2017a). Therefore, it is challenging for wildlife tourism providers who *do* provide ethical, conservation-oriented activities to differentiate themselves to tourists. Given the enormous economic clout of the tourist industry, many governments are reluctant to regulate wildlife tourism, and even when regulation is present, it is often not adequately enforced (Moorhouse et al., 2017a; WAP, 2017).

Trends

Currently, the market for captive wildlife attractions is projected to increase. Promotional materials for selfie safaris and similar captive attractions tend to naturalise proximity to wildlife and to greenwash its implications; these venues promote touching or holding wildlife as safe, fun, desirable and culturally authentic (Rizzolo, 2017). The use of captive wildlife as a resource for humans is reinforced by a market environmentalism approach to wildlife tourism that prioritises economic revenue over wildlife well-being (Belicia & Islam, 2018). Although close proximity to wildlife is desirable for many tourists (Rizzolo, 2017), Asian tourists are particularly interested in being close to and touching wildlife, and less likely than other tourists to prefer viewing animals in the wild (Moscardo & Saltzer, 2004). The attitudes and behaviours of Asian tourists will substantially shape the future of wildlife tourism. China is the world's largest outbound tourist market; it is double the size of the next largest market (the United States) and continues to grow (Boniface et al., 2016; Moorhouse et al., 2017b).

Selfie safaris are influenced not only by tourism market trends, but also by a combination of technological, sociological and environmental factors. Although the phenomenon of self-portraits is a long-standing one, the use of selfies has rapidly increased in the digital age, facilitated by technological advancements in mobile phones and access to social media platforms such as Instagram and Facebook (Lim, 2016). From 2012 to 2014 alone, the number of selfies increased by 900 times (Souza et al., 2015). These technological advancements have rapidly increased the reach of selfies; users can now easily exchange photographs with a global audience. Social media platforms allow people to have access to celebrity 'influencers'; when celebrities post photographs of themselves holding endangered animals, this can exponentially increase demand for selfie safaris due to the volume of their social media followers (Kitson & Nekaris, 2017). The wildlife selfies of only five celebrities can generate up to 1 billion views on Instagram (WAP, 2017). Thus, the spread of wildlife selfies through social media can both glamorise this behaviour and amplify demand (Kitson & Nekaris, 2017). At the same time, consideration of animal welfare issues in tourism is increasing (Liu et al., 2004), although tourist behaviours often contradict their stated values (Moorhouse et al., 2017b).

Wildlife selfies are also driven by sociological factors such as wealth, culture and social expression and communication. Although some may conceptualise selfies as the epitome of individualism, selfies are fundamentally social; they are meant to be shared with and seen by others, and they reflect and are shaped by larger economic and social trends (Pearce & Moscardo, 2015). Research on the selfie culture is quick to note that simply pathologising all selfies as manifestations of narcissism is overly simplistic. In a neoliberal economic system, selfies are multifaceted; they are simultaneously mechanisms of self-expression, forms of social currency and commodities that can be used to promote capitalistic products and services (Iqani & Schroeder, 2016). In part, the content of selfies is driven by what prior tourists have viewed as desirable, a self-perpetuating process of representation known as the hermeneutic cycle (Pearce & Moscardo, 2015). However, tourist markets are fluid and can shift to accommodate new economic or social trends. For example, as China becomes wealthier, more Chinese citizens are able to afford travel, with dramatic effects on the wildlife tourism market (Li et al., 2011). Increased wealth also allows international wildlife attractions to become more accessible and may be associated with a shift in preference away from domestic destinations (Balmford et al., 2009).

Finally, the choice to create and share selfies is embedded in cultural norms. Selfies tend to be more prevalent in countries characterised by gender equality (Souza et al., 2015). This may be due to the preference for this form of photography among young women; one study of slow loris selfies on Instagram found that 84% of these selfies were posted by women (Kitson & Nekaris, 2017). Selfies are more frequent in countries where people have a strong sense of belonging to the local community and less common in countries where citizens trust each other and have a sense of control over their lives (Souza et al., 2015). Cultural factors impact not only preferences and segmentation in wildlife tourism markets, but also how tourists respond to conservation messaging meant to discourage participation in harmful activities such as selfie safaris (Moorhouse *et al.*, 2007b).

Selfie safaris are also driven by environmental trends such as wildlife characteristics and infrastructure. Participation in wildlife tourism is influenced by the presence of desirable species, or species that are rare, valuable and/or 'cute' (Reynolds & Braithwaite, 2001). Given the increase in species threatened with extinction (Barnosky *et al.*, 2011), it is likely that many of the charismatic megafauna (e.g. elephants, tigers, primates) used in selfie safaris will become rarer over time. Species that are particularly sensitive to tourist impacts due to their biological characteristics (such as sloths; Carder *et al.*, 2018) may become threatened at an accelerated rate; while rarity can increase the appeal of wildlife species, it also makes it more difficult to access them.

Another relevant environmental driver is the development of infrastructure, such as roads and tourist facilities. Infrastructure both provides access to species and attracts more generalised or novice tourists (Duffus & Dearden, 1990). For example, the airport in Manaus, Brazil, can accommodate international aircraft as well as access to wildlife through paved roads and sizable riverboats; this infrastructure will likely contribute to the growth of an already booming wildlife selfie industry in this region (WAP, 2017). Since novice wildlife tourists often have different wildlife viewing preferences than more specialised tourists (Martin, 1997), infrastructure can also alter the tourism market of a locale towards more high-profile species. As a venue grows in popularity, it can attract participants who have less dedication to the wildlife species and can create a tourism market where operators focus on maximising visitor volume (e.g. the number of people taking selfies with a tiger) rather than the value of the activity (Bentz *et al.*, 2016).

Areas of Uncertainty

There are three main areas of uncertainty that will impact the future(s) of selfie safaris. First, will tourists change their behaviour to select wildlife tourism attractions that have beneficial environmental and animal welfare impacts? Given that governmental regulation of wildlife tourism is limited, tourist demand has a substantial impact on the type of attractions that tour operators provide. Priming, or providing information on tourism impacts prior to the selection of an attraction, can motivate tourists to choose wildlife attractions that have positive environmental and animal welfare impacts, although the benefits of priming are smaller for Chinese tourists than for Western tourists (Moorhouse *et al.*, 2017).

Secondly, will the social media platforms that facilitate the sharing of problematic wildlife selfies limit or ban these photographs? Although

Instagram, one of the most prolific selfie platforms, has introduced a new alert system to draw users' attention to behaviour that harms wildlife (Daly, 2017b), it remains to be seen whether this will serve as an effective deterrent. It remains unclear whether other sites that drive participation in selfie safaris, such as Tripadvisor and Facebook, have the will and/or tools to prevent the spread of these detrimental photographs.

Third, will the wildlife species used in selfie safaris become rarer? Many of the species that feature heavily in problematic selfies are endangered or threatened (WAP, 2017). If these animals become rarer, which is a distinct possibility given the looming sixth mass extinction (Barnosky et al., 2011), then demand for selfies with these species may actually increase, as rarity tends to increase tourist appeal and numbers (Booth et al., 2011).

Various outcomes for these three key uncertainties yield seven scenarios. The eighth combination is not elaborated further because it is neither consistent nor plausible that, if tourists don't engage in responsible tourism and social media does not regulate, the species used in selfie safaris will remain stable, given the severe environmental and animal welfare impacts of this form of wildlife tourism. I begin by delineating the most optimistic (utopia) and most pessimistic (dystopia) scenarios, and then discuss the other potential futures.

The Future(s) of Selfie Safaris

Utopia

Imagine that both social media platforms and individual tourists act to protect wildlife. A celebrity visits Thailand and poses for a selfie with a tiger at an attraction called Tiger Kingdom, and the photo goes viral. A month later, undercover footage is released of tigers at this venue being beaten, drugged and suffering from untreated illnesses and wounds. There are particularly horrendous photos of one particular tiger, nicknamed Stripes by the media, at this facility, and it is revealed that Stripes died at Tiger Kingdom. People begin sharing other footage of animal cruelty that they gathered while on vacation. Outrage is swift and widespread, amplified by social media. When people post this form of selfie, friends dislike the posts and post negative comments. Social pressure mounts and so does moral outrage. Other celebrities and members of the public convince Instagram to ban wildlife selfies. Driven by its desire to seem socially responsible to its primarily millennial users, Instagram develops a tool to detect and remove problematic wildlife selfies, and other social media platforms follow suit. Tourists begin booking more naturalistic forms of wildlife tourism, such as visits to national parks and sanctuaries where they maintain a respectful distance from the animals. The hashtag #wild trends on Instagram; the more remote and wild the tourism venue, the higher the social prestige.

When these naturalistic venues cannot keep up with demand, the tourism industry pressures governments to set aside more land and resources for this type of tourism. The governments of Thailand, Brazil and other hotspots of wildlife tourism prioritise nature-based tourism that allows wildlife to experience wild or semi-wild conditions. Countries in Asia and Latin America begin to experience economic booms from this type of naturalistic wildlife tourism, similar to the safari industry in parts of Africa. This further incentivises governments to set aside land for this purpose, to curb poaching and the illegal wildlife trade in order to protect their wildlife, and to prioritise animal welfare and conservation in order to attract and sustain tourists. As a result of these initiatives, populations of tigers and other endangered animals previously used in selfie safaris rebound and stabilise.

Dystopia

Now consider a very different type of scenario. In this future, neither tourists nor social media act to mitigate the harm of wildlife selfies. Several influential musicians, models and actors post photographs of themselves holding endangered species. These selfies become some of the most liked posts on Instagram, and collectively reach 2 billion views. Demand for selfies with already endangered species increases. Many of the species used for these selfies, such as sloths, don't fare well in their captive venues, becoming ill or dying after a short period of time. As a result, poaching pressures increase on wild populations already taxed by deforestation and habitat loss. Wild populations of sloths, slow lorises, tigers, elephants and other animals used as selfie props continue to drop. The increased rarity of these animals only escalates demand, which is met by poaching wild animals or breeding genetically compromised individuals in poor conditions. These venues attract tourists interested in a quick selfie rather than a prolonged trek to see wildlife; this leads to increased infrastructure, including roads, bridges and buildings, near these venues. In turn, this development further drives deforestation and the wild populations of these species plummet towards extinction.

Beyond these two extremes, there are a variety of other potential futures including:

Environmental decline from other sources

Tourist demand for selfie safaris is reduced and social media platforms stem the spread of problematic wildlife selfies. However, these actions are not enough to stabilise wildlife populations that are under threat from other factors such as habitat loss, climate change and the illegal wildlife trade. However, the reduction in selfie safaris slows the rate of species decline and relieves the suffering of thousands of animals previously kept in poor conditions.

Tourist actions are too little too late

While tourists shift to more naturalistic forms of wildlife tourism, social media platforms do not intervene to ban harmful wildlife selfies. Only a small percentage of visitors post these photographs, but some of these tourists are influential on social media and their posts reach millions of users. Although the wildlife tourism market diversifies and begins to offer more naturalistic options such as visits to wildlife sanctuaries, selfie safaris continue to flourish as a niche market for the wealthy.

Tourists save the day

In the absence of social media regulation, changes in the tourism market are sufficient to reduce selfie safaris and stabilise wildlife populations. Tourist outrage grows to the point that tourism operators and travel sites alter how they conduct business. Consider a scenario in which disturbing undercover footage from a selfie safari venue goes viral. Tourists leave a flood of negative reviews on Tripadvisor, Facebook and other travel-oriented sites. The venue is temporarily closed while the government conducts an investigation, and Tripadvisor marks this facility as closed. New tourists are not made aware of this venue, and demand dries up.

Social media mitigates harm

Tourists continue to participate in problematic wildlife selfies, but social media platforms mitigate the spread of these images. Consider if social media sites such as Instagram and Facebook classified problematic wildlife selfies as images of animal cruelty and worked to remove them from their sites. Now there are fewer opportunities to see the wildlife selfies of others (both friends and celebrities) and to share these images. Tourists looking for social recognition find other sites and activities to post about; those looking for 'extreme' selfies take selfies while sky-diving, rock-climbing or scuba-diving rather than posing with exotic animals.

Instagram falls short

Tourists don't choose more responsible wildlife tourism attractions, social media platforms do regulate and the species used for captive wildlife selfies still decrease in numbers. As tourism increases in countries like China, the wildlife tourism sector evolves. Facebook and Instagram are not available in China, so their efforts to prevent the spread of harmful wildlife selfies has little impact on Chinese residents. Suppose that, driven by a pre-existing preference for close proximity to wildlife, wildlife selfies become popular on Chinese social media sites such as Huaban. Chinese residents assume that these captive wildlife sites are effectively regulated. Demand for wild animals to fill these sites continues to grow.

Conclusion

There are numerous ethical, management and social implications of the rise of selfie safaris. As the scenarios presented in this chapter demonstrate, if no action is taken by either tourists or social media platforms, the harms of problematic wildlife selfies can accelerate, posing a danger to both sustainable tourism and biodiversity. The scenarios also illustrate the social nature of selfies. This form of photography is about social recognition and self-expression within a social context. While disrupting the spread of these images is one way to mitigate the ill-treatment of wildlife, another mechanism is to use social media to create change; the same social pressure that leads people to post detrimental wildlife selfies could be harnessed to encourage conservation-oriented behaviour.

Finally, the severe and adverse impacts of wildlife selfies contradict the traditional conceptualisation of live animal encounters as non-consumptive and ethically non-problematic. In planning for the future(s) of wildlife tourism, we must recognise that live animal encounters are heterogeneous. Posing with captive or baited wildlife and visiting a wildlife sanctuary will have vastly different impacts on the future of endangered species and wildlife tourism. The optimism of scenario planning lies in its ability to forecast multiple futures that can be altered through present-day action (Yeoman & Postma, 2014). The scenarios developed in this chapter can inform a new model of non-consumptive wildlife tourism that recognises the current and future harms of selfie safaris, and in doing so yields a more sustainable future for wildlife.

References

Amer, M., Daim, T.U. and Jetter, A. (2013) A review of scenario planning. *Futures* 46, 23–40.

Balmford, A., Beresford, J., Green, J., Naidoo, R., Walpole, M. and Manica, A. (2009) A global perspective on trends in nature-based tourism. *PLoS Biology* 7 (6), e1000144.

Barnosky, A.D., Matzke, N., Tomiya, S., Wogan, G.O., Swartz, B., Quental, T.B., Marshall, C., McGuire, J.L., Lindsey, E.L., Maguire, K.C., Mersey, B. and Ferrer, E.A. (2011) Has the Earth's sixth mass extinction already arrived? *Nature* 471 (7336), 51–57.

Belicia, T. and Islam, M. (2018) Towards a decommodified wildlife tourism: Why market environmentalism is not enough for conservation. *Societies* 8 (3), 59–74.

Bentz, J., Lopes, F., Calado, H. and Dearden, P. (2016) Managing marine wildlife tourism activities: Analysis of motivations and specialization levels of divers and whale watchers. *Tourism Management Perspectives* 18, 74–83.

Boniface, B., Cooper, C. and Cooper, R. (2016) *Worldwide Destinations: The Geography of Travel and Tourism*. London: Routledge.

Booth, J.E., Gaston, K.J., Evans, K.L. and Armsworth, P.R. (2011) The value of species rarity in biodiversity recreation: A birdwatching example. *Biological Conservation* 144 (11), 2728–2732.

Carder, G., Plese, T., Machado, F., Paterson, S., Matthews, N., McAnea, L. and D'Cruze, N. (2018) The impact of 'selfie' tourism on the behaviour and welfare of brown-throated three-toed sloths. *Animals* 8 (11), 216–228.

Daly, N. (2017a) Special report: The Amazon is a new frontier for deadly wildlife tourism. *National Geographic*, 3 October, np.
Daly, N. (2017b) Instagram fights animal abuse with new alert system. *National Geographic*, 4 December, np.
D'Cruze, N., Machado, F.C., Matthews, N., Balaskas, M., Carder, G., Richardson, V. and Vieto, R. (2017) A review of wildlife ecotourism in Manaus, Brazil. *Nature Conservation* 22, 1–16.
Duffus, D.A. and Dearden, P. (1990) Non-consumptive wildlife-oriented recreation: A conceptual framework. *Biological Conservation* 53 (3), 213–231.
Duinker, P.N. and Greig, L.A. (2007) Scenario analysis in environmental impact assessment: Improving explorations of the future. *Environmental Impact Assessment Review* 27 (3), 206–219.
Fennell, D.A. (2012) *Tourism and Animal Ethics*. London: Routledge.
Iqani, M. and Schroeder, J.E. (2016) # selfie: Digital self-portraits as commodity form and consumption practice. *Consumption Markets & Culture* 19 (5), 405–415.
Kitson, H. and Nekaris, K.A.I. (2017) Instagram-fuelled illegal slow loris trade uncovered in Marmaris, Turkey. *Oryx* 51 (3), 394.
Li, X.R., Lai, C., Harrill, R., Kline, S. and Wang, L. (2011) When east meets west: An exploratory study on Chinese outbound tourists' travel expectations. *Tourism Management* 32 (4), 741–749.
Lim, W.M. (2016) Understanding the selfie phenomenon: Current insights and future research directions. *European Journal of Marketing* 50 (9/10), 1773–1788.
Liu, Y., Zhang, W. and Tang, X.D. (2004) The prospect of wildlife tourism. *Journal of Forestry Research* 15 (3), 243–245.
Martin, S.R. (1997) Specialization and differences in setting preferences among wildlife viewers. *Human Dimensions of Wildlife* 2 (1), 1–18.
Moorhouse, T.P., Dahlsjö, C.A., Baker, S.E., D'Cruze, N.C. and Macdonald, D.W. (2015) The customer isn't always right—conservation and animal welfare implications of the increasing demand for wildlife tourism. *PloS One* 10 (10), e0138939.
Moorhouse, T.P., D'Cruze, N.C. and Macdonald, D.W. (2017a) Unethical use of wildlife in tourism: What's the problem, who is responsible, and what can be done? *Journal of Sustainable Tourism* 25 (4), 505–516.
Moorhouse, T.P., D'Cruze, N.C. and Macdonald, D.W. (2017b) The effect of priming, nationality and greenwashing on preferences for wildlife tourist attractions. *Global Ecology and Conservation* 12, 188–203.
Moscardo, G. and Saltzer, R. (2004) Understanding wildlife tourism markets. In K. Higginbottom (ed.) *Wildlife Tourism: Impacts, Management, and Planning* (pp. 167–186). Altona: Common Ground Publishing.
Mutalib, A. (2018) The photo frenzy phenomenon: How a single snap can affect wildlife populations. *Biodiversity* 19 (3–4), 237–239.
Nekaris, K.A.I., Campbell, N., Coggins, T.G., Rode, E.J. and Nijman, V. (2013) Tickled to death: Analysing public perceptions of 'cute' videos of threatened species (slow lorises–*Nycticebus* spp.) on Web 2.0 Sites. *PloS One* 8 (7), e69215.
Newsome, D. and Rodger, K. (2013) Wildlife tourism. In A. Holden and D. Fennell (eds) *The Routledge Handbook of Tourism and the Environment* (pp. 345–358). Oxford Routledge.
Newsome, D., Dowling, R.K. and Moore, S.A. (2005) *Wildlife Tourism*. Clevedon: Channel View Publications.
Page, S.J., Yeoman, I., Connell, J. and Greenwood, C. (2010) Scenario planning as a tool to understand uncertainty in tourism: The example of transport and tourism in Scotland in 2025. *Current Issues in Tourism* 13 (2), 99–137.
Pearce, J. and Moscardo, G. (2015) Social Representations of Tourist Selfies: New Challenges for Sustainable Tourism. Paper presented at the BEST EN Think Tank XV:

The Environment–People Nexus in Sustainable Tourism: Finding the Balance, Kruger National Park, 17–21 June.

Reynolds, P.C. and Braithwaite, D. (2001) Towards a conceptual framework for wildlife tourism. *Tourism Management* 22 (1), 31–42.

Rizzolo, J.B. (2017) Exploring the sociology of wildlife tourism, global risks, and crime. In M. Gore (ed.) *Conservation Criminology: The Nexus of Crime, Risk and Natural Resources* (pp. 133–156). New York: Wiley-Blackwell.

Rizzolo, J.B. (in preparation) Wildlife tourism and consumption.

Robertson, M. and Yeoman, I. (2014) Signals and signposts of the future: Literary festival consumption in 2050. *Tourism Recreation Research* 39 (3), 321–342.

Schmidt-Burbach, J., Ronfot, D. and Srisangiam, R. (2015) Asian elephant (*Elephas maximus*), pig-tailed macaque (*Macaca nemestrina*) and tiger (*Panthera tigris*) populations at tourism venues in Thailand and aspects of their welfare. *PloS One* 10 (9), e0139092.

Schoemaker, P.J. (1993) Multiple scenario development: Its conceptual and behavioral foundation. *Strategic Management Journal* 14 (3), 193–213.

Shackley, M.L. (1996) *Wildlife Tourism*. London: International Thomson Business Press.

Souza, F., de Las Casas, D., Flores, V., Youn, S., Cha, M., Quercia, D. and Almeida, V. (2015) Dawn of the Selfie Era: The Whos, Wheres, and Hows of Selfies on Instagram. Paper presented at the Proceedings of the 2015 ACM Conference on Online Social Networks, Palo Alto, CA, 2–3 November.

Tisdell, C. and Wilson, C. (2012) *Nature-Based Tourism and Conservation*. Northampton: Edward Elgar Publishing.

World Animal Protection (WAP) (2017) *A Close Up on Cruelty: The Harmful Impact of Wildlife Selfies in the Amazon*. New York: World Animal Protection.

Yeoman, I. and Postma, A. (2014) Developing an ontological framework for tourism futures. *Tourism Recreation Research* 39 (3), 299–304.

6 The Future of Captive Wildlife: Useful and Enjoyable for Animals and Visitors?

Ronda J. Green

Some of my earliest memories include looking up in awe at restlessly pacing lions and tigers, feeding peanuts to monkeys and elephants and gradually getting to know the ring-tailed lemurs, wapitis, currasows and other zoo creatures and dreaming about exploring Africa, the Amazon and other wild places. As I started to realise that the animals did not always seem happy, I dreamed of starting a zoo where I would go collecting in these wonderful places, then release all the animal after six months and collect new ones. It made sense to my six-year-old mind, and I read every wildlife book I could find. I no longer like the kind of zoo that it used to be, and get far more satisfaction from watching wild elephants splashing through an African river or orangutans feeding on fruits in a Bornean rainforest than any zoo, but it played a major role in prompting my interest in animal behaviour, wildlife conservation and my eventual career as a zoologist.

It is obvious to anyone who has visited zoos around the world that there is a spectrum of zoos, from those that perform admirably to others that are atrocious. Many recent posts on social media call for a total ban on zoos, either because many animals lead boring lives and suffer mistreatment, or from a feeling that no animal should ever be held captive. Zoo proponents argue that modern zoos offer essential services for conservation, research and education and staff are very attentive to animal welfare. Still, many animals throughout the world are regrettably kept purely for the entertainment of tourists and the profit of owners, without respecting the animals' needs. Keeping popular animals in small enclosures or permanently chained, or forcing them to perform unnatural tricks, has been justly condemned by many. Most cynical of all perhaps are institutions that attract animal-loving visitors to interact with animals in a supposed rehabilitation programme which is actually breeding them for canned hunting operations. This chapter imagines a future that is fundamentally different from this reality.

Tourism facilities that keep wild animals (zoos, aquaria, wildlife parks, etc.) remain a popular family pastime and a wildlife experience for tourists who do not have the time or physical stamina to seek animals in the wild. As income from visitors is often essential for conservation and research projects as well as keeping animals healthy, zoos and other captive wildlife operations must satisfy or, where possible, delight these visitors. Catering to the needs and desires of both humans and other animals is not always straightforward, and we need to pose the following questions:

- Why keep wild animals captive?
- What are the basic needs of wild animals in captivity?
- How can we help wild animals enjoy their lives in captivity?
- What do visitors need, want and expect?

The answers to these and related questions can help us visualise how zoos and other captive facilities of the future could truly assist wildlife as well as attract enough paying visitors to keep afloat financially. After answering the aforementioned questions and reflecting on how to balance the animals' and the tourists' needs, this chapter presents a futuristic scenario and comments on the related knowledge needs (Figure 6.1).

Why Keep Wild Animals Captive?

Nowadays a feature of many captive animal operations, conservation breeding has many guidelines and regulations. Although there have been many failures (e.g. oribi in South Africa and woma pythons in Australia quickly succumbing to predators upon release: Grey-Ross et al., 2010; Moseby et al., 2015), there have been numerous success stories, such as the European bison in Eastern Poland (Pucek, 2004) and the orange-bellied parrot of Australia (DELWP, 2019). Many problems exist, especially regarding feral predators (e.g. bilbies: Enoch, 2019) and lack of habitat for safe release (e.g. swamp turtle: Laschon, 2016), but pre-release training about predators can work (Ross et al., 2019), and it would seem better that endangered species persist in well-run captive facilities awaiting possible rehabilitation for their descendants rather than disappear altogether from our planet.

Another reason for keeping wild animals captive concerns animals that have been rescued for welfare reasons from various situations but cannot be released back into the wild because of physical condition, e.g. lame, too 'humanised' to recognise its own species, likely to be attacked by conspecifics. Similarly, a life in captivity can be the only option for animals rescued from zones of civil unrest, e.g. Jane Goodall's Chimpanzee Eden in South Africa (https://www.chimpeden.com).

Figure 6.1 Being up close to animals can be exciting for visitors. This panda at Adelaide Zoo has a roomy enclosure where he can still get some privacy among rocks and vegetation when he wishes (Photographer: R. Green)

Captive animals also provide an opportunity for intensive research on social behaviour, foraging choices, physiology, potential for artificial insemination and other aspects of their biology. Research into these fields can contribute to gaining important knowledge about the animals and how to better care for them. The education of visitors is an important role of many keepers of captive animals, and if done well, for example through knowledgeable guides with good communication skills and the innovative and effective use of new technologies, it has the potential to promote the appreciation and understanding of many animals by thousands of visitors.

What Are the Basic Needs of Wild Animals in Captivity?

Humans, including staff and visitors, choose to be in zoos. Other animals don't. Humans who keep animals captive for whatever reason thus have a responsibility to provide at least the basics of adequate food, water and shelter against extreme temperatures. The 'five freedoms' (e.g. Webster, 1994), first developed to improve conditions for farm animals, have also been used extensively as guidelines for captive wildlife around the world: freedom from (1) hunger and malnutrition; (2) discomfort and exposure; (3) pain, injury and disease; (4) fear and distress; as well as (5) freedom to express normal behaviour.

Providing these freedoms does not guarantee a fulfilling, enjoyable life, so this approach has now largely been superseded by the five domains: (1) nutrition, (2) environment, (3) health, (4) behaviour and (5) mental state, emphasising not just the elimination of negative conditions but also the provision of positive features (Mellor & Beausoleil, 2015). Mellor (2016) argues that instead of concentrating on the elimination of suffering, we should strive towards the positive goal of 'a life worth living' for all animals, and where possible a 'good life'. Similarly, Dawkins (2004) indicates that beyond the needs that keep animals alive, there are behavioural 'needs' when an animal strongly desires something it cannot have, such as migratory birds with an urge to fly during the season when their wild conspecifics are doing so.

Some provisions are obvious: space for young ungulates and carnivores to run around and play, structures for monkeys and other arboreal creatures to climb (Figure 6.2), soil suitable for burrowing animals, conspecific companions for social species. Some are not so clear-cut, and we cannot always generalise from one species to another related one – tigers for instance tend to be more energetic than lions, and chimpanzees more sociable than orangutans. Many years ago, the Austrian ethologist Konrad Lorenz (1961) also pointed out that public sympathy towards different species can be misleading. For example, zoo visitors might condemn

Figure 6.2 Ring-tailed lemurs at Western Plains Zoo, Dubbo, New South Wales, enjoy a three-dimensional enclosure with plenty of vegetation

the confinement of raptors while not realising the abject boredom of more inquisitive, active and playful parrots and cockatoos in small cages.

One common and often important measure of welfare is the level of stress. While mild stress can sometimes be unimportant or even positive (e.g. arousing an animal from time to time into healthy exercise or exploratory behaviour or sexual urges leading to copulation), frequent stresses or severely stressful situations can be deleterious to an animal's well-being. Morgan and Troburg (2007) point out some of the difficulties in determining stressful situations. Firstly, many animals, from giraffes to hummingbirds, have senses beyond our own, e.g. infra-sound, ultrasound, certain frequencies of flickering lights and many levels and kinds of scent. Moreover, some animals may also feel a lack of opportunity not obvious to humans: e.g. wombats not allowed sufficient time for seeking food (Hogan & Tribe, 2006) or klipspringers without high platforms from which to survey their surroundings (Rose et al., 2017).

Human interaction is a potential source of stress very relevant to wildlife tourism, as many tourists seek close interactions with animals. Baird et al. (2016) found the stress levels of armadillos, hedgehogs and hawks handled by keepers and others for exhibits or educational purposes to be significantly associated with handling frequency and duration, enclosure size and substrate, but again we cannot automatically generalise to other species, or even to conspecifics with different background experience and temperament.

Recent scientific advancements have influenced wildlife tourism practices, moving the sector towards better animal welfare (e.g. Borges de Lima & Green, 2017; Carr & Broom, 2018; Fennell, 2013; Shani & Pizam, 2008). Whitham and Wielebnowski's (2013) research involving brain mechanisms associated with motivation and emotion is potentially useful for physiological and biological markers of high levels of welfare. Rose et al. (2017) identify major risk factors and potential solutions to the development of stereotypic behaviour in apes and monkeys. Clark (2011) proposes a new framework for great ape enrichment, inspired by a century of cognitive research. With regard to the assessment of animal welfare, Wolfensohn et al. (2018) review some of the methods and challenges. Brando and Buchanan-Smith (2017) point to the complexities of the different needs of species and individuals, including individual reactions in social contexts, and propose a welfare assessment tool based on 14 criteria. The Association of British Travel Agents (ABTA, 2015) offers extensive guidelines for good practice in wildlife tourism, both captive and free-ranging, that are being adopted worldwide.

How Can We Help Wild Animals Enjoy Their Lives in Captivity?

Dawkins (2004) argues that instead of asking the simple but sometimes ambiguous question of what the animals need, we should be asking two relevant but separate questions: (1) Are the animals healthy? and (2) Do the animals have what they want? She gives examples of experiments to – as she put it in her presentation at the International Ethology Conference in 1987 (Wisconsin) – 'ask the animals' what they want, instead of leaping to conclusions based on preconceived ideas: e.g. chickens have been experimentally found to choose a more complex pen over a bare one, and to be more willing to venture outdoors if their yard includes trees. Dawkins (2004) recommends experiments with choices to be conducted in places where the animals live, such as within farms or zoos, rather than situations set up in a laboratory, and predicts that this will become an increasingly important feature of improving animal welfare.

Knowledge of a species' ecology and behaviour in the wild can also suggest ways of providing enrichment. Reading *et al.* (2013) demonstrate the added value of behavioural enrichment for animals that are to be rehabilitated into the wild, increasing their chances of success after release. Chimpanzees at Taronga Zoo (Sydney) are provided with artificial termite mounds and a daily supply of stems to strip to size and 'fish' for mustard (https://taronga.org.au/news/2018-07-11/turning-heads-tools). Bryan *et al.* (2017) found the behavioural repertoire of tamarins in a large free-range exhibit allowed a behavioural repertoire approaching that of wild individuals.

Shepherdson (1994) provides several examples of enrichment that mimic nature and improve welfare: bears with hidden food and objects to manipulate; a walrus provided with realistic foraging opportunities; and a kinkajou provided with food that required exploratory and manipulatory behaviour. A question that can be raised is: How 'natural' do enrichment experiences need to be to enhance animal welfare? The animal's perception may be different to our own automatic reaction. Dawkins (2004) cautions that 'it is not the naturalness or otherwise of the behaviour that is critical, but the extent to which it can, in any given instance, be linked to either physical health or what the animal does or does not want'. Wolfensohn *et al.* (2018: 6) also caution that 'environments that fully mimic "the wild" are not necessarily better for welfare... and providing for optimal captive zoo animal welfare should provide some aspects of the wild environment (such as opportunity for foraging, exploration and choice) and withhold some of the stressors (such as presence of predators)'. Veasey (2017) suggests that the level of welfare of captive animal populations protected from the key physical stressors of the wild might at times be higher than they would routinely experience in wild populations.

Grandin (2018) also indicates that animal welfare and being 'natural' are two different things, and speaks of the obvious enjoyment of a captive polar bear playing with a ball. At 'An Evening with Jane Goodall' in Brisbane in 2017, Dr Goodall was asked what she thought about captive chimpanzees using computers. She said they love it, as boredom can be one of the worst aspects of captivity. Orangutans have been observed to enjoy iPads and painting materials (Webber *et al.*, 2017). Social animals, such as many primates, wolves and many herd animals and flocking birds need to socialise. Apart from the obvious bonds between mother and infant or mated pairs, they can form long-lasting friendships (Seyfarth & Cheney, 2013) and it could be distressing to them to be separated. Some individuals also simply do not like each other and forcing them into lifelong close proximity may be another source of stress.

Play has a number of functions, including the practice of motor skills for climbing, running or manipulating objects, exercise beneficial to health and learning social skills. Byrne (2015) notes that it has become more acceptable in scientific circles to acknowledge that while these underlying functions undoubtedly exist and have been selected for by evolution, the individual animal plays because it enjoys doing so, it has fun. The provision of play items for enrichment has become increasingly common in the past couple of decades, providing entertainment for both the animals and the onlookers.

What Do Tourists Need, Want and Expect?

Visitors have basic needs of health and safety: e.g. they should never have the opportunity for deliberate or unintentional contact with genuinely dangerous animals, and visitors who contact animals should be provided with antibacterial handwash. Beyond that, there are the questions of what they want or expect, which managers must take note of, since they generally depend on visitors for income. People generally visit zoos to see wildlife, and while some are prepared to sit and wait for animals to appear or accept the reasons they may not be visible, others can get upset if some animals are not immediately visible. Many also quickly get bored if the animal is not actively doing anything interesting or they can't see it up close or interact with it by patting, feeding or riding.

Many zoo visitors have favourite animals plus species they don't enjoy seeing. The major reasons tourists at Jersey Zoo cited certain wildlife exhibits as their least favourite were that the animals were 'not visible', 'stuff of nightmares', 'bad smell', 'boring/unexciting', 'ugly' and 'not active' (Carr, 2016). Carr (2016) suggests that identifying why certain animals are unattractive to visitors 'is just the first step in a process that should have as its end point an increase in zoo visitor appreciation for animals that have been identified as visitors' least favourite' via innovative planning of enclosures, viewing methods and education techniques. Although entertainment appears to be the primary motivation for visiting a zoo, some visitors express interest in conservation and education. Roe and McConney's (2015) study suggests that over 70% of visitors come to a zoo expecting to learn something. Smith *et al.* (2012) examined the reactions of visitors to repeated conservation messages, and concluded that there was little objection to multiple conservation messages or to the same message repeated multiple times.

Balancing the Needs of Wildlife and Tourists

Based on the aforementioned considerations, some reflections can be done on the possibility of balancing the animals' and the tourists' needs. With regard to physical place, it can be noted that large enclosures, usually preferred for the animals, may necessitate long walking distances for visitors. To face such challenge, some zoos use safari-style buses or allow private vehicle access or provide bicycles for hire.

As noted in the previous sections, animals need seclusion from visitors at times. Cameras under sheltering shrubs, two-way mirrors and peep-holes still allow a feeling of closeness of which the animal is happily unaware. Webber *et al.* (2017) explore the use of interactive technological devices to achieve a feeling of closeness.

Where close encounters and interactions are allowed, animal welfare can be enhanced by using only individuals that appear to enjoy or at least tolerate interaction with humans, and not using the same animal every day or for more than a couple of hours on any particular day. Photos of visitors with animals may be justified by their role in benefiting other animals: for instance at the Currumbin Wildlife Sanctuary, Gold Coast, Australia, paid-for photos of visitors with koalas raise money for their wildlife hospital, which tends their own animals and also many injured and orphaned ones brought in by members of the public. Many institutions also allow people to pat or hold animals during educational activities, in the hope that the heightened emotional response to close contact will reinforce the educational value of the guide's talks (Webber *et al.*, 2017). Nygren and Ojalammi (2018) reviewed research articles on conservation education in zoos, and concluded that it is most effective when the animals are active or charismatic, there is some kind of contact with the animal, post-visit materials or activities support the learning aims, visitors are already conservation minded and the setting is interactive and naturalistic.

Most studies on human–wildlife interactions have investigated possible negative impacts on animals. Sade (2013) found only a few studies on positive effects, and suggests that more research is needed. Some animals actually seek attention from humans, and not just begging for food. This emerges also from some of my personal experience. For example, once a young chimpanzee at a zoo invited me to scratch her back through the bars, and I obliged. A dolphin in an aquarium tossed a ball to me and indicated he wanted me to throw it back, which again I did. I have also play-wrestled a half-grown jaguar that a vet released from its night-cage – not something I would recommend for visitors unaccustomed to playful and powerful youngsters, but both the big cub and I enjoyed our game.

Many tourists search for eye contact with the animals and this can be experienced without compromising the welfare of the animals. One hard-to-please tourist I travelled with briefly in Kruger National Park was entranced when a leopard walked close to our vehicle. 'She looked at me! She looked right at me! I think I'm in love!' After being introduced to a flying fox at a wildlife expo in a district where this species is unpopular for various reasons, an elderly farmer said later 'looking into that animal's eyes was a life-changer for me'.

An Imaginary Zoo of the Future

It is 2050: I'm a tourist and I'm intrigued by the promotion of a zoo that advertises not only an exciting experience but also its commitment to providing enjoyable lifestyles for its animals, a chance to see endangered species that I'm not likely to see elsewhere, and its many conservation achievements. Its website provides many details of wildlife ecology and behaviour, as well as information on the species it has assisted through conservation breeding and the release of animals that have now started successfully breeding in the wild, the decrease in poaching where the zoo has helped fund ranger programmes and the increase in animal populations where the zoo has funded habitat restoration along with fencing and overpasses to protect from feral predators and straying onto busy roads. They also state on their website that they opened the zoo in a regional area both to enable more space for the animals and to contribute to the local economy, choosing a location where entry is close to a railway station for visitors without cars. I decide to visit this zoo!

Even at the entrance, the education and entertainment begin, with artistic murals and large-screen videos while waiting for tickets, including conservation issues affecting each of the animals inside, what the zoo is doing to help and success stories so far. We can then choose from a range of free brochures or downloadable apps: puzzles and quizzes to solve while exploring displays, links to details on critically endangered species, behaviours to watch for, how to donate to various conservation projects, etc. One that I pick up explains the design of the exhibits: how they shield prey animals from the scent, sight and sound of predators, and afford secluded refuges to all animals but with cameras so visitors can still see them. There is a sudden gasp from the crowd entering the zoo, as an elephant appears nearby, flapping its enormous ears and throwing up its trunk. Then we relax as it fades: it is a hologram – one of many at the zoo, featuring animals popular with visitors but which need many hectares that are better used for the conservation breeding of highly endangered species.

I first head to the African section. Looking down to one side, I can see African wild dogs in a wooded enclosure which adjoins the path for 100 metres but stretches for half a kilometre away from us, giving these restless animals ample room to run swiftly across the grass and explore among woodpiles and boulders. Potential prey animals housed on the opposite side of the path are unable to see the dogs. A holographic herd of wildebeest now gallop through the wild dog enclosure, just long enough to excite the dogs into further activity. My app tells me why wildebeest and other animals make their long migrations and the problems of routes now being blocked in parts of Africa. A young boy reads a sign and pulls a lever that causes a large robust toy to drop and dangle from a tree for a minute, while the dogs leap, grab and tug at it until it springs back into the branches, and I now see other toys dotted inconspicuously through the trees: fun for humans and dogs alike. By the time a small group of young adult dogs is ready for release into the wild, they will be fit and healthy and ready to tackle challenges in their new home.

In the Australian section, I walk down a ramp to a subterranean tunnel to view animals such as bilbies, burrowing crayfish and blind snakes, parts of whose burrows are lit by infra-red light and can be viewed through glass, while videos on the walls show some of their behaviour. The bilbies can also be viewed aboveground in special night tours offered by the zoo.

A spiral staircase (near an elevator for the less fit) leads to a tunnel through a forest canopy in the Asian section, where gibbons rush past us, brachiating through the branches. The 300 metre enclosure allows them impressive speed. Others feed on fruits above and below us, including some imported from Indonesian forests, and play with robust toys that blend well with their surroundings, sometimes stopping to look at a large video screen supplied for their entertainment, showing monkeys, birds, lizards and occasionally wild gibbons. A colourful sign tells me that due to much research on rehabilitation methods the progeny of these gibbons has been shipped in small groups of socially bonded individuals and successfully introduced to protected forests in their ancestral homeland of Indonesia. High above me, not accessible by the gibbons or humans other than zoo staff, is a hollow in which an endangered hornbill is nesting. I can't see her directly but can view her on the live-screening video screen in front of me, sometimes being fed by her mate. After descending almost to ground level, I can choose one of two ways to reach the exit – continue down the ramp or brachiate my way above another ramp with a rubbery surface to cushion falls, and my app tells me that only apes and humans (not monkeys) can brachiate, and gibbons are the fastest at it. I'm tempted to try but today it is crowded with enthusiastic youngsters.

Close encounters are permitted with some of the more inquisitive animals, especially individuals which are not to be reintroduced to the wild. I enjoy entertaining a lemur and a parrot by manipulating various toys and adopting strange postures, and wear a virtual reality headset that seems to take me right into the rhino enclosure with the keeper who enters at feeding time. I also get a feeling for the life of mahogany gliders by gliding below a strong cable through the branches, and see if I can build a bowerbird bower as well as the bird can do it. I role-play a potoroo in a computer game that introduces me to fascinating behaviours and their ecological significance, so as I go around the corner I immediately feel a kind of bond with the potoroos housed in the next enclosure, and a heightened understanding of their world.

I leave at closing time knowing I have only seen a fraction of what this zoo has to offer, and understand why so many visitors buy yearly passes and also enter quizzes on wildlife behaviour and conservation to win vouchers for special behind-the-scenes tours.

Future Research Needed

Even the best zoos can be progressively improved as we learn more about the various species, including our own. Crane (2007) laments the lack of scientific information in many quarters, causing regulators of zoo exhibit standards to rely too heavily on 'anecdotal evidence, expert opinions and extrapolations from scientific evidence about other animals'. For example, we need more knowledge about animals and stress, in particular in relation to interaction with humans. Under what circumstances do animals become stressed, tolerate or actively enjoy interaction with humans, including strangers? Baird et al. (2016) recommend future investigation into thresholds or weekly limits to handling animals, individual animal personalities and factors such as crowd size and animal choice contributing to the tolerance of handling. The effects of human participants on wildlife viewing, close encounters and interactions with wildlife have been the subject of recent study (Ballantyne et al., 2018), but there is much scope to further refine our understanding. Interestingly, Carlstead and Shepherdson (1994) indicate that most research on zoo animal behaviour is conducted during daylight hours, but that much of importance happens during the night, deserving extra study.

Zoos can and do conduct their own research, but each zoo will normally have only a few animals of each species, often insufficient for statistical analysis and valid conclusions. Carlstead and Shepherdson (1994: 447) predict that 'multi-variate multi-institutional behavioural research in zoos will play an increasingly important role in the successful captive propagation of many species by closely examining relationships between environmental variables and reproductive potential of individual animals'. Whitham and Wielebnowski (2013: 247) also speak of future welfare science involving 'trans-disciplinary, multi-institutional studies and epidemiological approaches' related to welfare indicators. The more we understand, the better zoos can develop to fulfil important functions.

I no longer enjoy the bar-and-cement kind of zoo that first introduced me to the wonders of wildlife. I would love to see children of the future captivated and wanting to protect animals they contact in zoos that contribute significantly to conservation and research and also keep up with the latest research for making their venues not just free of suffering but actually as enjoyable for the animals as for the visitors. This is one of my visions for the future of captive wildlife: how about you?

References

ABTA (2015) *Global Welfare Guidance for Animals in Tourism*. London: ABTA.

Baird, B.A., Kuhar, C.W., Lukas, K.E., Amendolagine, L.A., Fullera, G.A., Nemeta, J., Willisb, M.A. and Schook, M.W. (2016) Program animal welfare: Using behavioural and physiological measures to assess the well-being of animals used for education programs in zoos. *Applied Animal Behaviour Science* 176, 150–162.

Ballantyne, R., Hughes, K., Lee, J., Packer, J. and Sneddon, J. (2018) Visitors' values and environmental learning outcomes at wildlife attractions: Implications for interpretive practice. *Tourism Management* 64, 190–201.

Borges de Lima, I. and Green, R.J. (eds) (2017) *Wildlife Tourism, Environmental Learning and Ethical Encounters: Ecological and Conservation Aspects*. Cham: Springer.

Brando, S. and Buchanan-Smith, H.M. (2017) The 24/7 approach to promoting optimal welfare for captive wild animals. *Behavioural Processes* 156, 83–95. doi: 10.1016/j.beproc.2017.09.010

Bryan, K., Bremner-Harrison, S., Price, E. and Wormell, D. (2017) The impact of exhibit type on behaviour of caged and free-ranging tamarins. *Applied Animal Behaviour Science* 193, 77–86.

Byrne, R.W. (2015) The what as well as the why of animal fun. *Current Biology* 25 (91), 2–4.

Carlstead, K. and Shepherdson, D. (1994) Effects of environmental enrichment on reproduction. *Zoo Biology* 13, 447–458.

Carr, N. (2016) An analysis of zoo visitors' favourite and least favourite animals. *Tourism Management Perspectives* 20, 70–76.

Carr, N. and Broom, D. (eds) (2018) *Tourism and Animal Welfare*. Wallingford: CABI.

Clark, F.E. (2011) Great ape cognition and captive care: Can cognitive challenges enhance well-being? *Applied Animal Behaviour Science* 135, 1–12.

Crane, M. (2007) Without the wisdom of Solomon or his ring: Setting standards for exhibited animals in New South Wales. *Journal of Veterinary Behavior* 2, 223–229.

Dawkins, M.S. (2004) Using behaviour to assess animal welfare. *Animal Welfare* 13, S3–7.

DELWP (2019) Orange-bellied parrot *Neophema chrysogaster*. Threatened Species Fact Sheets. Department of Environment, Land, Water and Planning. Victoria State Government. See https://www.environment.vic.gov.au/conserving-threatened-species/threatened-species-fact-sheets/orange-bellied-parrot (accessed 17 August 2020).

Enoch, L. (2019) Bilbies get new home just in time for Easter. Department of the Environment and the Great Barrier Reef, Queensland Government. See http://statements.qld.gov.au/Statement/2019/4/18/bilbies-get-new-home-just-in-time-for-easter (accessed 17 August 2020).

Fennell, D.A. (2013) Tourism and animal welfare. *Tourism Recreation Research* 38 (3), 325–340.

Grandin, T. (2018) My reflections on understanding animal emotions for improving the life of animals in zoos. *Journal of Applied Animal Welfare Science* 21 (1), 12–22. doi: 10.1080/10888705.2018.1513843

Grey-Ross, R., Downs, C.T. and Kirkman, K. (2010) An assessment of illegal hunting on farmland in KwaZulu-Natal, South Africa: Implications for oribi (*Ourebia ourebi*) conservation. *South African Journal of Wildlife Research* 40, 43–52.

Hogan, L.A. and Tribe, A. (2007) Prevalence and cause of stereotypic behaviour in common wombats (*Vombatus ursinus*) residing in Australian zoos. *Applied Animal Behaviour Science* 105, 180–191.

Laschon, E. (2016) Western Swamp Toortoise faces slow recovery despite Perth Zoo breeding success. ABC News, Sydney. https://www.abc.net.au/news/2016-04-29/western-swamp-tortoise-faces-long-road-to-recovery-perth-zoo/7371910 (accessed 17 August 2020).

Lorenz, K. (1961) *King Solomon's Ring* (M. Kerr Wilson, trans.). London: Methuen.

Mellor, D.J. (2016) Updating animal welfare thinking: Moving beyond the 'Five Freedoms' towards 'A Life Worth Living'. *Animals* 6, 21. doi: 10.3390/ani6030021

Mellor, D.J. and Beausoleil, N.J. (2015) Extending the 'Five Domains' model for animal welfare assessment to incorporate positive welfare states. *Animal Welfare* 24, 241–253.

Morgan, K.N. and Tromborg C.T. (2007) Sources of stress in captivity. *Applied Animal Behaviour Science* 102, 262–302.

Moseby, K.E., Peacock, D.E. and Read, J.L. (2015) Catastrophic cat predation: A call for predator profiling in wildlife protection programs. *Biological Conservation* 191, 331–340.

Nygren, N.V. and Ojalammi, S. (2018) Conservation education in zoos: A literature review. *Trace: Finnish Journal for Human-Animal Studies* 4. doi: https://doi.org/10.23984/fjhas.66540

Pucek, G. (2004) European bison: Status survey and conservation plan. IUCN/SSC Bison Specialist Group. See https://ibs.bialowieza.pl/g2/pdf/1436.pdf (accessed 17 August 2020).

Reading, R.P., Miller, B. and Shepherdson, D. (2013) The value of enrichment to reintroduction success. *Zoo Biology* 32, 332–341.

Roe, K. and McConney, A. (2015) Do zoo visitors come to learn? An internationally comparative, mixed-methods study. *Environmental Education Research* 21 (6), 865–884. doi: 10.1080/13504622.2014.940282

Rose, P.E., Nash, S.M. and Riley, L.N. (2017) To pace or not to pace? A review of what abnormal repetitive behavior tells us about zoo animal management. *Journal of Veterinary Behavior Clinical Applications and Research* 20, 11–21. doi: 10.1016/j.jveb.2017.02.007

Ross, A.K., Letnic, M., Blumstein, D.T. and Moseby, K.E. (2019) Reversing the effects of evolutionary prey naiveté through controlled predator exposure. *Journal of Applied Ecology* 56, 1761–1769. doi: 10.1111/1365-2664.13406

Sade, C. (2013) Visitor effects on zoo animals. *The Plymouth Student Scientist* 6 (1), 423–433.

Seyfarth, R.M. and Cheney, D.L. (2013) Social relationships, social cognition, and the evolution of mind in primates. In R.J. Nelson and S. Mizumori (eds) *Comprehensive Handbook of Psychology, Volume 3: Biological Psychology and Neuroscience* (pp. 574–594). New York: John Wiley & Sons.

Shani, A. and Pizam, A. (2008) Towards an ethical framework for animal-based attractions. *International Journal of Contemporary Hospitality Management* 20 (6), 679–693.

Shepherdson, D. (1994) The role of environmental enrichment in the captive breeding and reintroduction of endangered species. In P.J.S. Olney, G.M. Mace and A.T.C. Feistner (eds) *Creative Conservation* (pp. 168–177). Dordrecht: Springer.

Smith, L.D.G., Curtis, J., Mair, J. and van Dijk, P.A. (2012) Requests for zoo visitors to undertake pro-wildlife behaviour: How many is too many? *Tourism Management* 331, 502–510.

Veasey, J.S. (2017) In pursuit of peak animal welfare; the need to prioritize the meaningful over the measurable. *Zoo Biology* 36 (6), 413–425.

Webber, S., Carter, M., Smith, W. and Vetere, F. (2017) Interactive technology and human–animal encounters at the zoo. *International Journal of Human-Computer Studies* 98, 150–168.

Webster, J. (1994) Assessment of animal welfare: The five freedoms. In J. Webster (ed.) *Animal Welfare: A Cool Eye Towards Eden* (pp. 10–14). Oxford: Blackwell Science.

Whitham, J.C. and Wielebnowski, N. (2013) New directions for zoo animal welfare science. *Applied Animal Behaviour Science* 147 (2013), 247–260.

Wolfensohn, S., Shotton, J., Bowley, H., Davies, S., Thompson, S. and Justice, W.S.N. (2018) Assessment of welfare in zoo animals: Towards optimum quality of life. *Animals* 8 (7), 110. doi: 10.3390/ani8070110

7 Promises and Pitfalls in the Future of Sustainable Wildlife Interpretation

Gianna Moscardo

> To excite curiosity, to open a person's mind – there is challenge for anyone who seeks to communicate ideas.
> Tilden, 1967

Freeman Tilden is often described as the father of interpretation, which is the process of designing visitor experiences around educational activities that help them experience natural protected areas. It is Tilden's approach to interpretation and visitor experience that underpins this chapter, exploring how new technologies might influence the future of interpretation in wildlife tourism experiences (WTE). Although there is no evidence that Tilden travelled outside North America before his death in 1980, based on his writings, we can imagine his reaction to a night wildlife viewing tour in the Wet Tropics World Heritage listed rainforests of the far north-eastern part of Australia. This fictional account gives us some insights into the challenges and tensions that currently exist in wildlife visitor experiences.

So, let's imagine Freeman on a visit to Northern Australia keen to experience both the natural environment of this exotic location and the nature of Australian interpretation. He joins Bob, a local guide, and a group of other tourists for a night wildlife spotlighting tour in the rainforest, a classic type of wildlife interpretation and a typical WTE in this location. Freeman, like the other tourists, chose the tour because the brochure suggested that Bob had years of experience as a guide and that this was a tourist's best chance of seeing a Lumholtz tree kangaroo. The Lumholtz tree kangaroo is a rare animal adapted for the niche taken by monkeys in many other rainforests. It exists only in the rainforests of the Atherton Tablelands, the setting for our tour, and is difficult for visitors to see because it is nocturnal and lives most of its life high in the rainforest canopy eating leaves. Bob begins the tour by outlining the safety messages that apply and how the visitors need to act so that they do not negatively impact the creatures they are likely to encounter on the tour, especially by using only the filtered light torches they are given and to switch off camera flashes so as not to disturb the nocturnal animals. Freeman has heard this information several times before and, despite his background as an enthusiastic naturalist, he finds it dull. He wonders to himself, 'If I'm finding this dull, how are my fellow tourists reacting?'

Bob heads off with enthusiasm, listing all the creatures he had found for his previous tour group, two nights before. Bob is cheerful and friendly and very good at spotting and pointing out various creatures including several Boyd's forest dragons (an arboreal species of lizard usually about 30 cm in length that is a relic species linking the rainforest back to the Jurassic era) and a common brushtail possum, and identifying the distinctive call of the Lesser Sooty Owl (which sounds like a falling Second World War bomb). Bob is full of facts about each of these animals and encourages and helps the visitors to take photographs of the dragons and the possum. Bob is in a hurry to get to the spots where he has previously seen a very large amethystine python (one of the world's largest snakes) and a tree kangaroo. He and several of the tourists talk about ticking these off their lists of must-see creatures. In his rush, he passes by several other smaller creatures, well actually small for creatures in general, but very large for insects (such as the Hercules moth with a wingspan of up to 30 cm and the giant burrowing cockroach which can grow up to 8 cm in length). He dismisses them as not being the real prize and moves quickly on. In his haste, he also doesn't notice that one of his tourists, someone Freeman noted seems not to understand English, has been taking photographs of the wildlife they have seen with a flash after Bob has moved on and has often moved off the path breaking foliage to get the best shot.

Sadly for everyone the python and the tree kangaroo are not there and after another hour of searching Bob brings the tour to a close. Bob is sad he didn't find the two creatures he was searching for, but the group is very grateful for his general enthusiasm. Freeman is also disappointed that he didn't get to see a Lumholtz tree kangaroo, but he is even more disappointed that Bob wasn't able to take the opportunity of being in such an extraordinary natural environment and the wildlife that the group did see, to weave a story about the importance of this environment and to connect this place to the visitors. Freeman sees a wasted opportunity to light some sparks among the visitors, to give them some guidance on how they can look for wildlife themselves and what they can do beyond the tour to help these creatures.

This narrative vignette (Hughes & Huby, 2012) highlights several key challenges or tensions in the use of interpretation for WTEs. The first is balancing a focus on charismatic individual species versus the larger environment that they live in. Bob's attention was so focused on finding the tree kangaroo that he missed many opportunities to link the wildlife he did find to its environment and to provide some insights into why this environment was special and important. The second challenge is managing visitor expectations and experiences in environments where the wildlife may be difficult to find or see. The scenario also demonstrates the tensions between providing visitors with positive experiences, which are often created around getting close to and/or interacting with the wildlife and learning about the wildlife (Moscardo, 2006), and managing the negative impacts of the encounter through limiting access and proximity.

These challenges reflect a wider tension within interpretation between the different functions that people believe interpretation should fulfil and in assumptions about the importance of interpretation in enhancing the sustainability of tourist experiences. A review of the definitions and descriptions of heritage interpretation suggests several functions including visitor management and visitor experience enhancement or substitution (Moscardo, 2015). The tension lies between interpretation focused on the first function of managing, controlling and limiting tourism impacts on the heritage and interpretation focused on the second function of creating and enhancing visitor experiences. There is an argument that connects the two with the proposal that a positive visitor experience encourages visitors to adopt a conservation or care ethic that supports minimal impact behaviour messages both at the site and beyond the site. Tilden's original view that interpreters should light a spark by presenting just enough access and information that visitors could create their own personally meaningful experience that would lead to greater reflection on issues linked to sustainability, is an example of this argument linking experience to management. The rapid and substantial increase in tourism and visitation to heritage sites subsequent to Tilden's time working in the field has, however, resulted in a much greater focus on site hardening and interpretive messages controlling, limiting and directing visitor behaviour (Ablett & Dyer, 2009; Moscardo, 2017a). Some have argued that this focus on visitor control and minimising immediate negative impacts at the site detracts from both the visitor experience (Moscardo, 2017a) and the ability to make any meaningful change to visitor attitudes or actions beyond the site. Bertella (2019) argues that sustainability in wildlife tourism requires an approach based on animal ethics and that sustainable WTE needs to offer experiences that are less exploitative and that encourage visitors to take the perspective of the wildlife itself. This search for WTE that encourages more sustainable and ethical tourist behaviour seems unlikely to be fulfilled by the increasing focus on visitor control. Given that a core argument for the inclusion of interpretation in wildlife tourism is to encourage learning about wildlife and its habitats, support wildlife conservation and increase visitor awareness of animal ethics, it could be proposed that encouraging more mindful wildlife tourists is critical to moving towards greater sustainability in wildlife tourism.

This chapter will argue that emerging mobile information and communication technologies (ICT) offer opportunities to address these challenges. These trends combined with the changing global tourist profile may, however, also present problems for the management of visitor–wildlife interactions. This chapter will take a futures thinking approach to explore the likely impact of these disruptions on interpretation in WTE. While the focus for the examples and scenarios will be on non-captive WTE offered in natural environments, many of the arguments and conclusions will apply to other wildlife-based tourism settings.

Taking a Futures Thinking Approach

Futures thinking seeks to examine possible changes in key processes or phenomena in order to understand how to prepare for them and how to move towards more desirable outcomes (Varum *et al.*, 2011). A core element of futures thinking is the development and use of scenarios or stories about what the future might be (Burnam-Fink, 2015). Amer *et al.* (2012: 25) reviewed the use of scenarios in future thinking and argued that scenario building is important to help overcome limited thinking and to assist people to move beyond their 'traditional operational and conceptual comfort zone'. In keeping with these futures thinking traditions, this chapter is based on an intuitive logic approach to scenario building and adapts the steps outlined by Inayatullah (2008) for building and using scenarios for futures thinking. Table 7.1. briefly introduces these steps, which are then described in more detail in the following sections and applied to the challenges facing interpretation to support sustainable WTE.

Table 7.1 Steps in developing futures scenarios

Inayatullah's analytical steps		Approach for developing futures scenarios for interpretation in wildlife experiences
Mapping the present	→	Identifying key elements of interpretation in WTE
Deeper systems analysis	→	Analysing systems supporting interpretation in WTE
Anticipating disruption	→	Exploring ICT trends and changing global tourist profiles
Create alternative scenarios	→	Future scenarios for interpretation in WTE
Transformation	→	Conclusions about sustainability of WTE

Identifying Key Elements of Interpretation in Wildlife Tourism Experiences

Table 7.2 provides a summary of the characteristics that have been consistently shown to increase visitor satisfaction in reviews of tourist experiences in general, and WTE in particular. As can be seen, the characteristics of multisensory features, being interactive and engaging,

Table 7.2 Characteristics of satisfying tourist experiences

Characteristics	Tourist experiences in general	Wildlife tourist experiences
Multisensory	X	X
Interactive and engaging	X	X
Personally relevant	X	–
Educational or informative	X	X
Emotional	X	X
Novel (rare or previously unseen species of wildlife)	X	X
Immersive or themed	X	–
Close to, or easy to see wildlife	–	X
Variety	X	X
Clear orientation and activity structure	X	X
Specific animal features – infants, perceived intelligence, cultural symbols	–	X

Sources: Apps *et al.*, 2017; Ballantyne *et al.*, 2011; Curtin, 2010; Moscardo, 2006, 2017a; Schmitt & Zarantello, 2013; Zhong *et al.*, 2017.

educational, novel, varied, emotional and with a clear orientation and activity structure apply to both types of experience. Table 7.2 also lists additional elements consistently linked to satisfactory wildlife tourism and these are mostly focused on the wildlife elements.

What is also noteworthy about these characteristics of a satisfactory experience are the elements that are linked to deeper cognitive or mental reflection on the experience. Moscardo (2006, 2013) connected wildlife experiences to mindfulness theory from cognitive psychology. Mindfulness theory (Langer, 2014) is a well-established dual processing theory focusing on explaining attention, learning and behaviour change in everyday social situations and has been used by thousands of researchers across a range of fields and topics including education (cf. Yeh *et al.*, 2018), sustainability action (Wang *et al.*, 2019) and tourism (Moscardo, 2019). Dual processing is a foundation concept in psychology (Strack & Deutsch, 2015) and argues that in any given situation a set of individual factors and setting characteristics combine to encourage one of two cognitive states. The first is a deep, systematic or mindful state in which an individual pays detailed attention to the immediate setting, seeking important information and making decisions about new routines of behaviour (Langer, 2014). The alternative is a shallow, heuristic or mindless state where individuals pay minimal attention to their immediate setting and use established routines, scripts or habits to guide their actions (Langer, 2014).

The important feature of this dual processing concept for the present discussion is that mindfulness is associated with a greater likelihood of positive emotional and evaluative responses to an experience, learning, the ability to take an alternative perspective on a situation and attitude and behaviour change (Langer, 2014). All of these elements are important goals of interpretation and sustainability in WTE. Mindfulness is encouraged when experiences:

- are able to be personalised and visitors are given control so that they can build meaningful links to their own experiences and values;
- are varied, immersive and multisensory;
- can provide visitors with alternative perspectives on a topic;
- are built around coherent and emotionally engaging stories (Kang & Gretzel, 2012; Moscardo, 2013).

It is worth noting that none of these principles was apparent in the vignette at the start of this chapter but do appear in Table 7.2.

Systems Analysis of WTE

A key feature of futures thinking is understanding the nature of the system that is being analysed. McCool *et al.* (2015) argue that tourism research has not engaged in very much systems analysis and this has limited its contribution to improving the management of tourism in general, as well as to improving tourism and sustainability. Fortunately for the present study, Hughes and Moscardo (2019) offer a systems analysis of managing tourist experiences that can be adapted to examining interpretation and wildlife-based experiences. This systems analysis examined the nature of tourist experiences from the perspective of the tourists, focusing on the decisions tourists have to make and the types of problems they need to solve. It identified five touchpoints where tourists and suppliers come into direct contact, including:

- making *choices* or decisions about what to do and where to go within a destination or attraction;
- customising these experiences to meet personal needs and interests;
- seeking to build *connections* to each other and to the place being visited;
- *co-creating* personally meaningful experiences by linking information and stories to their activities;
- *complying* with site or attraction management directives designed to keep tourists safe and to minimise their negative impacts on the site and its occupants or users.

The first four of these touchpoints are processes that tourists use to try and create positive and mindful experiences, while the fifth is about how tourism managers seek to control and limit tourist impacts. This division again mirrors the tensions that have already been identified and the five touchpoints provide a way to focus attention on how disruptions may change the nature of interpretation in WTE.

Exploring ICT Trends and Changing Global Tourist Profiles

Traditional approaches to futures studies identify the need to examine changes and trends in the areas of energy, ecology and economy, geo-political shifts, sociocultural evolution and science and technology in order to develop broad global scenarios (Saritas & Smith, 2011). This specific study on interpretation and wildlife tourism futures chose to focus on only two of these major areas – science and technology and sociocultural change. Even more specifically, the analysis concentrated on three related technology trends, the rise of mobile computing and communication linked to increasing interaction through social media platforms and supported by the rise of the internet of things, especially through near-field sensing, and one aspect of sociocultural change, the rise of new international tourist markets.

Recent statistics indicate that tourists rely heavily on their mobile communication devices and their social media networks to make travel decisions, to communicate with both fellow travellers and social groups beyond the destination and to record and share their experiences (Wang et al., 2016). Mobile devices are used to find travel information, to make travel bookings, to decide on activity participation, for orientation, to take photographs and to share experiences (Lalicic & Weismayer, 2016). Many of these functions are achieved through social media platforms (Choe et al., 2017). With the introduction of near-field sensing and location-based services as part of the internet of things, it is also possible to connect tourists more directly to their immediate environment with electronic beacons sending information directly to mobile communication devices as tourists pass by (Manyika et al., 2013). Hughes and Moscardo (2019) argue that these new ICTs can enhance each of these touchpoints. For example, the ability to access detailed, up-to-date information and advice from social media reference groups allows tourists to make more informed and personally relevant choices. The ability to search the internet and social media platforms and use multiple travel apps also allows these choices to be customised to personal interests. Tourists are not reliant just on the information provided by the site interpreters or managers. This ability to immediately access and communicate with others, both at and beyond the site, enhances social connections and interactions. The rise of virtual and augmented reality (AR and VR) and gamification also allows tourists to participate in more engaging, responsive and immersive activities and to co-create experiences. All these processes contribute to mindfulness and thus should support greater attention paid to management safety and minimal impact messages. Compliance with these messages may also be supported directly by these new technologies. AR and VR can offer substitute experiences that direct visitors away from fragile sites. Mobile applications linked to location-based devices can provide onsite interpretation without the need to install signs and interpretive panels. Such technology can also monitor and remind visitors of safety and minimal impact issues through phone messages or alarms. These mobile location-based technologies can be seen as a type of surveillance that was not previously possible in many tourist locations, which may be sufficient to deter many visitors from engaging in inappropriate actions.

The widespread adoption and extensive use of mobile ICT can also be linked to the changing global profile of international tourists. The fastest-growing international tourist markets include China and India. China has come to dominate international tourism and is the largest source of international tourist departures with Chinese tourists accounting for 21% of all international tourism expenditure (UNWTO, 2018). Indian outbound tourism is much smaller than China, but growing very rapidly with the UNWTO predicting that there will be 50 million outbound Indian tourists in 2020 (UNWTO, 2018). While most tourism discussion of these new tourist markets has focused on their potential as markets for tourist businesses, there is increasing concern about how to manage both the volume of tourism from these countries and their onsite actions as they are very different tourists (Moscardo, 2017b). Initial discussions of wildlife interpretation for Chinese tourists notes a strong desire to buy animal products, to see wildlife presented in shows and to take photographs close to and/or holding wildlife, and a view that commercialisation of wildlife is a pathway to wildlife conservation (Cui *et al.*, 2012, 2017; Moscardo, 2017b). Such differences in attitudes and expectations are likely to change demand for WTE and create challenges for interpreters in many tourism settings.

Some Futures Scenarios for Wildlife Interpretation: Promises and Pitfalls

Hicks and Holden (1995) suggested four main categories of scenarios – more of the same, edge of disaster, technological fix and desirable or sustainable. To fit within space limitations, the present study will focus on two contrasting future scenarios – one that combines the technological fix and an aspirational or sustainable future, the promise, and one opposing option that combines more of the same with the edge of disaster, the pitfalls. Yeoman and Postma (2014) describe four different approaches to futures thinking that can be used in tourism research including one that seeks to describe utopias or dystopias which suggest different possible futures based on analysing social changes. Such scenarios are useful for identifying potential issues and challenges, as well as pathways to desirable goals (Yeoman & Postma, 2014). In the broader context, Ramirez *et al.* (2015) argue that the use of scenarios can challenge existing assumptions and open up new ways of thinking about both research and practice.

Future Scenario 1: The promise

Our fictional tourist, Freeman, joins Bob, a local guide, and a tour group for a night wildlife spotlighting tour in the rainforest. Freeman chose this tour based on recommendations on social media and because of its eco-certification confirmed through the sustainable tourism choices app on his mobile phone. Bob begins the tour at the Malanda Rainforest Interpretation Centre where a VR tree kangaroo, Kimberley, was created and installed as part of the exhibition in 2017. The VR experience gives Freeman insights into the biology of tree kangaroos, explains the things that threaten them and provides advice on how Freeman can be a responsible tourist and limit his negative impacts on tree kangaroos. The tourists learn some tips on how to spot tree kangaroos and get more information on some of the other species they are likely to encounter.

Before they leave the centre, they are reminded to download the tour app. The app begins with a short safety and minimal impact quiz game that they all have to complete before the experience begins. The app also automatically switches off the flash function on their cameras and will keep it off until they leave the rainforest tour area, and it switches on a torch with the correct filtered light to minimise their impacts on the wildlife. The app also allows them to customise the information they will receive to their personal language and individual interest preferences. Some choose the option of an educational game that encourages them to search for things in the rainforest and to use all their senses.

Bob is able to use his own version of the tour app. Small sensors placed along the trail monitor motion and the artificial intelligence inside these sensors can use the patterns of movement to identify the likely species that are in the area. It sends this information to Bob, allowing him to target his spotting more effectively. He finds several Boyd's forest dragons, a common brushtail possum, a giant burrowing cockroach and several Hercules moths. As Bob can more easily find and identify creatures, he is able to focus more attention on his guests. The same system that can identify animal movement can also detect when the tourists are on the path and so Bob is also warned when one of the group is lagging behind and stepping off the path. Bob moves back to check on this tourist, but the app has already politely reminded the tourist in his own language of the importance of staying on the path and he is hurrying to catch up with the group and apologising for his forgetfulness.

Unfortunately, the group has no luck finding a tree kangaroo and at the end of the trail Bob enables a final feature of the tour app, an AR option that allows the guests to see a virtual tree kangaroo in the setting, which is based on a real animal that Bob has encountered in the past. While watching this very realistic depiction, Bob weaves a story about this tree kangaroo, how it lives and the challenges it faces, asking the group to consider what the rainforest looks like to the tree kangaroo and how they might respond to the challenges it faces. The group is very happy with their experience and they explore the links from the tour app to rainforest conservation sites donating to various programmes and signing up for reminders about how they can help save these rainforests and the tree kangaroos when they get home.

Future Scenario 2: The pitfalls

Our target tourist, Freeman, and his teenage son, Ben, are with a group of tourists at a resort located adjacent to a rainforest in north-eastern Australia. Freeman, like the other tourists, chose this resort based on recommendations from his friends on social media. After looking at the social media posts with the most likes, Ben decides that he really wants to impress his friends at home by sharing a photo of himself with the rare Lumholtz tree kangaroo. The resort has a captive tree kangaroo, it's been cloned and born in captivity, which means the resort is not breaking any wildlife protection regulations. Guests can take photographs with the clone, but Ben wants a more impressive image, one taken in the actual rainforest. Ben finds out that the resort has developed an app that allows guests to take their own rainforest spotlighting tour, so Ben's group decides to use that to try and find a tree kangaroo. The resort, possibly illegally but it's difficult to know for sure as legislation has been slow to respond to these technologies, has hacked into the electronic monitoring devices that conservation agencies use to research and monitor the local tree kangaroo population. This means that the resort guests can use their app to track down the tree kangaroos. Although the guests can see where all the tree kangaroos are, they can't always get to them in the thick rainforest. Ben's group has found two kangaroos in locations the guests can easily get to. The camera flashes disorient and daze one of the tree kangaroos so much that one guest is able to climb and capture it. Ben is thrilled as he is able to get a photograph holding it. They put it back in the tree when everyone has their photo and they head back, posting their images to their social media. Later in the trip, they are asked to donate to a tree kangaroo conservation programme. Ben thinks to himself, it's not really necessary, the tree kangaroos were easy to find and don't really seem in danger. Besides, we can clone them so there's always that option.

Discussion and Conclusions

The deeper systems analysis conducted as part of this futures approach indicated that an important prerequisite for sustainable WTE are mindful visitors who can consider an alternative perspective on the situation, specifically who can think from the animal's perspective. Although, traditionally, interpreters have often assumed that their role and actions encouraged mindfulness, there is little systematic evidence that this is the case (Moscardo, 2014). This has been especially the case as increasing visitor pressures move management towards interpretation focused on controlling visitors. This control is increasingly challenged by the rapid growth in visitor numbers to many places, the changing profile of international visitors and easy access to multiple sources of information available through mobile access to ICT. Some would argue that this new technology not only creates pressure on the system but also offers some ways to enhance both the mindfulness and management of visitors in wildlife tourism.

The creation of the two futures scenarios offers a way to imagine how these opportunities and pressures might come into play in future WTEs. The positive utopian option highlights the ways in which interpreters and managers can use new technologies to nudge visitors into desirable actions and encourage them to be mindful. These new technologies could help light Tilden's spark by returning control over the cognitive and emotive dimensions of the experience to the visitor. The negative or dystopian scenario imagines the opposite future in which the technology rules unchecked. This use of two contrasting imaginative future scenarios offers insights into issues rarely considered in the literature, outlining the potential positive benefits of new technologies in tourism. In particular, it focuses attention on the dark side of pressures to remain connected to social media audiences, on the ways in which technologies to help tour operators find wildlife can threaten those wildlife and on the unexpected consequences of being able to use technology to partially adapt to the negative consequences of unsustainable action. This contrast leads us to the real power of imaginative futures thinking, implications and directions for changing current practices. The single greatest difference between the two scenarios is the underlying value placed on the wildlife. In the former, wildlife is seen as an entity valued for its existence, while in the latter, it is merely an object that exists to meet the needs of humans. This suggests that greater attention needs to be paid now to animal ethics if we wish to move towards more sustainable wildlife tourism in the future.

References

Ablett, P. and Dyer, P. (2009) Heritage and hermeneutics. *Current Issues in Tourism* 12 (3), 209–233.

Amer, M., Daim, T. and Jetter, A. (2013) A review of scenario planning. *Futures* 46, 23–40.

Apps, K., Dimmock, K., Lloyd, D. and Huveneers, C. (2017) Is there a place for education and interpretation in shark-based tourism? *Tourism Recreation Research* 42 (3), 327–343.

Ballantyne, R., Packer, J. and Sutherland, L. (2011) Visitors' memories of wildlife tourism. *Tourism Management* 32 (4), 770–779.
Bertella, G. (2019) Sustainability in wildlife tourism. *Tourism Review* 74 (2), 246–255.
Burnam-Fink, M. (2015) Creating narrative scenarios. *Futures* 70 48–55.
Choe, Y., Kim, J. and Fesenmaier, D. (2017) Use of social media across the trip experience. *Journal of Travel & Tourism Marketing* 34 (4), 431–443.
Cui, Q., Xu, H. and Wall, G. (2012) A cultural perspective on wildlife tourism in China. *Tourism Recreation Research* 37 (1), 27–36.
Cui, Q., Liao, X. and Xu, H. (2017) Tourist experience of nature in contemporary China. *Journal of Tourism & Cultural Change* 15 (3), 248–264.
Curtin, S. (2010) What makes for memorable wildlife encounters? *Journal of Ecotourism* 9 (2), 149–168.
Hicks, D. and Holden, C. (1995) *Visions of the Future*. London: Trentham Books.
Hughes, K. and Moscardo, G. (2019) ICT and the future of tourist management. *Journal of Tourism Futures* 5 (3), 228–240. doi 10.1108/JTF-12-2018-0072.
Hughes, R. and Huby, M. (2012) The construction and interpretation of vignettes in social research. *Social Work & Social Sciences Review* 11 (1), 36–51.
Inayatullah, S. (2008) Six pillars: Futures thinking for transforming. *Foresight* 10 (1), 4–21.
Kang, M. and Gretzel, U. (2012) Effects of podcast tours on tourist experiences in a national park. *Tourism Management* 33 (2), 440–455.
Lalicic, L. and Weismayer, C. (2016) The passionate use of mobiles phones among tourists. *Information Technology & Tourism* 16 (2), 153–173.
Langer, E.J. (2014) *Mindfulness, 25th Anniversary Edition*. Boston, MA: Merloyd Lawrence.
Manyika, J., Chiu, M., Bughin, J., Dobbs, R., Bisson, P. and Marrs, A. (2013) *Disruptive Technologies*. See https://www.mckinsey.com/~/media/McKinsey/Business%20Functions/McKinsey%20Digital/Our%20Insights/Disruptive%20technologies/MGI_Disruptive_technologies_Full_report_May2013.ashx (accessed February 2017).
McCool, S., Freimund, W. and Breen, C. (2015) Benefiting from complexity thinking. In G. Worboys, M. Lockwood, A. Kothari, S. Feary and I. Pulsford (eds) *Protected Area Governance and Management* (pp. 291–326). Canberra: ANU Press.
Moscardo, G. (2006) Is near enough good enough? Understanding and managing customer satisfaction with wildlife based tourism experiences. In E. Laws, B. Prideaux and G. Moscardo (eds) *Tourism and Hospitality Services Management* (pp. 38–53). London: CABI.
Moscardo, G. (2013) The role and management of non-captive wildlife in ecotourism. In R. Ballantyne and J. Packer (eds) *International Handbook on Ecotourism* (pp. 351–364). Cheltenham: Edward Elgar Publishing.
Moscardo, G. (2014) Interpretation and tourism. *International Journal of Culture, Tourism & Hospitality Research* 8 (4), 462–476.
Moscardo, G. (2015) Stories of people and places. In C.M. Hall, S. Gössling and D. Scott (eds) *The Routledge Handbook of Tourism and Sustainability* (pp. 294–304). London: Routledge.
Moscardo, G. (2017a) Critical reflections on the role of interpretation in visitor management. In J.N. Albrecht (ed.) *Visitor Management in Tourist Destinations* (pp. 170–187). Wallingford: CABI.
Moscardo, G. (2017b) Guidelines for Managing Mainland Chinese Tourists to National Parks 2017. Report. James Cook University, Townsville, QLD, Australia. See https://researchonline.jcu.edu.au/51786/ (accessed 15 May 2019).
Moscardo, G. (2019) Encouraging hospitality guest engagement in responsible action. *International Journal of Hospitality Management* 76, 61–69.
Ramirez, R., Mukherjee, M., Vezzoli, S. and Kramer, A.M. (2015) Scenarios as a scholarly methodology to produce 'interesting research'. *Futures* 71, 70–87.

Saritas, O. and Smith, J. (2011) The big picture–trends, drivers, wild cards, discontinuities and weak signals. *Futures* 43 (3), 292–312.

Schmitt, B. and Zarantonello, L. (2013) Consumer experience and experiential marketing. *Review of Marketing Research* 10, 25–61.

Strack, F. and Deutsch, R. (2015) The duality of everyday life: Dual-process and dual system models in social psychology. In M. Mikulincer, P.R. Shaver, E. Borgida and J.A. Bargh (eds) *APA Handbook of Personality and Social Psychology* (Vol. 1, pp. 891–927). Washington, DC: American Psychological Association.

Tilden, F. (1967) *Interpreting Our Heritage* (2nd edn). Chapel Hill, NC: University of North Carolina Press.

UNWTO (2018) World Tourism Barometer March/April 2018. See http://cf.cdn.unwto.org/sites/all/files/pdf/unwto_barom18_02_mar_apr_excerpt__0.pdf (accessed 15 May 2019).

Varum, A., Melo, C., Alvarenga, A. and Soeiro de Carvalho, P. (2011) Scenarios and possible futures for hospitality and tourism. *Foresight* 13 (1), 19–35.

Wang, D., Xiang, Z. and Fesenmaier, D. (2016) Smartphone use in everyday life and travel. *Journal of Travel Research* 55 (1), 52–63.

Wang, J., Geng, L., Schultz, P.W. and Zhou, K. (2019) Mindfulness increases the belief in climate change. *Environment & Behavior* 51 (1), 3–23.

Yeh, Y., Chang, H. and Chen, S. (2019) Mindful learning. *Computers & Education* 132, 63–75.

Yeoman, I. and Postma, A. (2014) Developing an ontological framework for tourism futures. *Tourism Recreation Research* 39 (3), 299–304.

Zhong, Y., Busser, J. and Baloglu, S. (2017) A model of memorable tourism experience. *Tourism Analysis* 22 (2), 201–217.

8 Interspecies Communication and Encounters with Orcas

Giovanna Bertella

Cetaceans are among the animal species that fascinate us the most. Throughout history and across cultures, cetaceans have featured in mythological stories. For example, whales appear in the myths of the Aivilik Eskimos, and dolphins appear frequently in classical Greek mythology (Carpenter, 1956; Williams, 2006). Cetaceans appear also in several world-famous stories from classic novels, such as Herman Melville's 1851 novel *Moby Dick*, and several pop-culture films, including the TV series *Flipper* from the 1960s and the *Free Willy* movie franchise (1993–2010). This fascination with cetaceans is particularly evident in the tourism industry: tourism involving whales, orcas and bottlenose dolphins in particular is a growing sector, as numerous tourists wish to see and be close to these animals (Higham *et al.*, 2014).

In controlled settings, tourists can experience various types of close encounter with cetaceans. Despite having been subject to quite severe criticism, aquaria that keep cetaceans, such as orcas and dolphins, in captivity remain common (Cater, 2010; Desmond, 1999; Lück & Jiang, 2007). Some of these facilities arrange in-water encounters with dolphins: tourists can swim close to the animals and, in some cases, touch and feed them. Such encounters tend to be viewed as extraordinary, memorable experiences by the participating tourists. Some tourists reportedly try to establish a connection of sorts with the animals, for example, through eye contact with some even going so far as to kiss them (Campos *et al.*, 2017; Curtin *et al.*, 2006).

In the wild, boat-based whale watching is a popular activity. In-water encounters with wild cetaceans can also occur. In the commercial context of tourism, these encounters are usually referred to as swimming-with programmes (SWPs) and are performed by snorkelers and divers (Convention on Migratory Species, 2017; Parsons & Brown, 2018; Parsons *et al.*, 2006; Wiener, 2013).

This study reflects on the future of tourism encounters with cetaceans. It is inspired by some of philosopher Jacques Derrida's reflections as well as ecofeminism's ideas concerning human–non-human animal relationships. It explores the possibility that cetaceans play a central and active role in the development and management of certain aspects of tourism, such as whale watching and aquaria. More precisely, it envisions the possibility that scientific advancements will allow humans and orcas to communicate with one another. This possibility is investigated in relation to three potential alternative futures, developed using fictional sources concerning cetaceans and real episodes and events that have occurred in Northern Norway's tourism destinations.

This chapter begins by presenting this study's philosophical and theoretical approach, and the method adopted with regard to the fictional scenarios. Next, these scenarios are described and discussed. The final part of this chapter reports three lessons that are relevant to current wildlife tourism practices and that can be learned from the imagined future scenarios, allowing us to reflect more deeply on our approach to wildlife.

The Philosophical Basis for the Inclusion of Animals in the Development and Management of Wildlife Tourism

The main philosophical source of inspiration for this study is the paper 'The Animal that Therefore I am (More to Follow)' by Jacques Derrida. In this paper, Derrida reflects on the ontology of non-human animals, the differences between humans and other animals and human–animal relations. His provocative thought concerns the perception of animals as subjects that are the objects of our attention and, at the same time, subjects that look upon and address us. Derrida illustrates this idea with a story about his cat observing him stepping out of the shower. In that moment, the man's and the animal's eyes met: the man felt almost ashamed and became fully aware of the fact that the animal was looking at him and had a specific point of view about him. This recognition triggered a mechanism and the man began to consider himself from the cat's perspective: a subject seen by another subject across the blurring boundary between species.

Derrida is not the only philosopher to have investigated the subjectivity of non-human animals and the possibility of mutual relations. For example, in their reflections on animal ethics, ecofeminists highlight the importance of recognising and respecting the individuality and dignity of each animal (Gaard, 1993). They argue in favour of establishing meaningful reciprocal relationships between humans and non-human animals, and, in addition, they comment on the moral obligation for humans to base such relations on the value of care (Gruen, 2015). Similarly, some biologists have considered and discussed the possibility of meaningful relationships between humans and wild animals. For example, Frans De Waal (2016) notes that the closer an animal species is to humans, the easier it is to identify some common ground for reciprocal understanding.

The aforementioned considerations are particularly relevant to wildlife tourism, as one of the few contexts in which people can experience wild animals and, consequently, reflect on their value – both in absolute terms and in relation to humanity – and establish relations (Gibson, 2010). Similarly, it can be argued that, from the animals' perspective, wildlife tourism offers opportunities to meet and learn about humans. This chapter reflects on the possibility that the animals that some tourists are so eager to meet might observe the tourists in return, thereby forming a precise idea of what humanity is and its value, in general and in relation to their animality. What might the animals' perspective on humanity be?

Cetaceans as Subjects Belonging to a Different Culture

Scientists have reached a relatively broad consensus on cetaceans having highly developed cognitive abilities and very rich emotional lives (Marino *et al.*, 2007). Cetology studies have long investigated the lives of whales, dolphins and porpoises, including their abilities, social structure and behaviour (Mann *et al.*, 2000). The results of such studies indicate that cetaceans are intelligent sentient beings. For example, bottlenose dolphins are reportedly able to learn, remember and innovate: they have a good understanding of 'symbolic representations of things and events (...), how things work or how to manipulate them (...), the activities, identities, and behaviors of others (...), of one's own image, behavior, and body parts' (Marino *et al.*, 2007: 970). With regard to social communication, sensory systems, including visual, tactile and acoustic, are found to be important, and differences have been identified among species as well as among those belonging to different pods (Deecke *et al.*, 2000; Pryor, 1990).

Remembering, learning, innovating and communicating are all processes that can be related to the existence of a culture, understood here as a set of information and behaviours shared among the members of a group of individuals through social learning (Rendell & Whitehead, 2001). Increasingly, evidence emerges indicating that cetaceans have their own culture, which is transmitted within and between generations (Whitehead *et al.*, 2008). This has also been observed in relation to the importance of including cetaceans' cultural dimensions in conservation strategies at policy level: 'it is important to consider the relevance of animal culture for scientific assessments and policy decision-making' (Brakes *et al.*, 2019: 3).

Further to these considerations, it might be reasonable and worthwhile to include recognition of the cetaceans that are objects of attention within the tourism sector as subjects with a distinctive culture. The perception of cetaceans as resources for humans and – in the tourism context – as mere attractions is hardly acceptable from an ethical standpoint. Their exploitation as resources, both in captivity and in the wild, may be viewed as a form of colonialism: a practice based on the use of power to gain some unilateral benefits (Ballengee-Morris, 2002; Montford, 2016). On the contrary, cetaceans might be reconceptualised as subjects entitled to active involvement in the development and management of those activities that affect their lives, tourism included.

Cetaceans: From Tourism Attractions to Co-Developers and Co-Managers of Wildlife Tourism

Collaboration in tourism is usually realised through communication and negotiation processes among various relevant actors, including tourism companies, destination management organisations (DMOs), public agencies at different levels, host communities, academia and non-governmental organisations (NGOs) (Bramwell & Lane, 2000; Phi & Dredge, 2019).

Collaboration among various stakeholders at different levels has also been discussed in relation to sustainable whale watching (Higham *et al.*, 2014). Although clearly important with regard to whale-watching experiences, the animals do not participate in any decisions concerning tourism: they are not directly involved in the development and management of the relevant tourism activities. When included, their perspective is represented by other actors, typically scientists or/and NGOs.

Bertella *et al.* (2019) argue that if animals could influence tourism development and management, wildlife tourism might differ considerably from its current state. This consideration relates to one of the major challenges in collaborative partnerships: the agreement on a shared vision and the possible existence of various values, interests and ways of operating (Leslie *et al.*, 2018; Scheyvens *et al.*, 2016). With regard to cetaceans, we may ask whether the animals' values, interests and ways of operating might be obscured by those of the tourism industry. This raises several questions: Do cetaceans value their freedom? Are they willing to give up their freedom for the sake of living in safer, albeit spaciously restricted, spaces? Are orcas happy to be trained by humans to perform for tourists? How would an orca describe happiness? Are cetaceans aware of the role they play in tourism? Do they have any understanding of what tourism is? Do the dolphins in aquaria like to be touched and kissed by tourists? To what extent do whales like or tolerate the proximity of whale-watching vessels, kayaks and snorkelers?

To explore questions of this nature, Bertella *et al.* (2019) develop a fictive dialogue between a tourist and a dolphin. During this dialogue, the animal makes explicit her views about encounters with snorkelers, reporting a range of thoughts and emotions provoked by these encounters. The authors conclude that the inclusion of animals as co-developers and co-managers of tourism activities may be problematic on several levels. Here, the animals' lack of experience and understanding of the human world, both in general and in the context of tourism, and humans' limited understanding of the animal world are viewed as major obstacles to such hypothetical collaboration. The fact that humans and cetaceans are unable to properly communicate with one another can be viewed as the key challenge to reciprocal understanding and possible collaboration.

This study imagines a future time when the obstacles to cetacean–human communication have been overcome, and asks what would happen if cetaceans, traditionally viewed as attractions or 'silent stakeholders' in the need for someone to represent their interest, could express their own perspective on tourism.

Method

This chapter adopts a creative approach to future scenario development. Some scholars argue that narratives can be used to describe and imagine the realities experienced by subjects that differ from us, and fictional novels and movies can be valuable sources of inspiration for such reflections (Parry & Johnson, 2007; Richardson & St. Pierre, 2005; Yeoman & Mars, 2012). This chapter presents four scenarios concerning the possible future developments of whale-watching tourism in Northern Norway. The first scenario occurs in 2050 and presents the background and the point of departure for three alternative 2055 scenarios. The name of the main character in the scenarios, Dr Brin, derives from David Brin, the author of a science fiction saga, *Uplift*, where humans have collaborative relations with dolphins.

The 2055 scenarios are inspired by:

- the 1986 movie *Star Trek IV: The Voyage Home*, in particular the representation of whales as particularly intelligent and empathic animals that can collaborate with humanity;
- the 1974 *The New York Times* article 'The Great Whale's Mistake' concerning the whale's perspective on humanity (Baker, 1974);
- People for the Ethical Treatment of Animals (PETA) advertisements concerning the captivity of orcas, in particular those representing humans detained in captivity (PETA, 2019).

Although it uses fiction and imaginary scenarios as its main sources of inspiration for envisioning these futures, this study is also based on some real episodes and events that have occurred in Northern Norway and which have raised concern among local people, including scientists, and the national authorities. These episodes and events are:

- the boom in the whale-watching sector since 2011 and the associated challenges in terms of uncontrolled growth (Bertella, 2017);

- the interference of oil activities in the lives of cetaceans living in the marine area of Northern Norway (Bertella & Vester, 2015);
- the increased risk of accidents due to the widespread and chaotic diffusion of recreational and tourist-oriented swim-with-orcas activities (Department of Fisheries, 2018).

The Futures of Whale Watching in Northern Norway

Background: 2011–2050

In 2011, Northern Norway became a world-famous whale-watching destination. From 2020, SWPs, the use of kayaks, stand-up paddles (SUPs) and jet-skis to view whales in close proximity became popular. The use of air and underwater drones also became widespread among tourists as well as local residents.

The growth in whale-watching activities ceased quite abruptly in 2025, when an accident occurred involving one kayaker and two snorkelers. The kayaker died, and the snorkelers were seriously injured. As a result of this incident, and on the basis of several coastguard reports regarding irresponsible behaviours on the part of some whale watchers, the government enforced several restrictive regulations. Whale watching from kayaks, SUPs and jet-skis was banned, and so were SWPs. Whale-watching activities for private individuals were forbidden. Tourist-oriented whale-watching activities organised by commercial actors were regulated with a licensing system that resulted in the selection of a limited number of operators that used green technology-based vessels. Such operators were also involved in an international research project investigating orca–human communication. This project was funded by local investors interested in exploring possibilities to develop the tourism sector in a more responsible and lucrative way.

In 2050, the orca–human communication research project reached an important milestone, when communication between humans and orcas was officially declared possible. Dr Brin, who has led the project since its inception, was enthusiastic. He began to envision the establishment of a frank dialogue with the animals and perhaps some collaborative relationships aimed at protecting the marine environment. Furthermore, the project's investors were interested in possible collaboration with the orcas and were thrilled to think about the new perspectives that collaborative relationships with the animals might yield with respect to the development of high-quality tourism products.

Alternative 2055 futures

Let us now imagine that the year is 2055. Dr Brin enters the conference room and is ready to inform the investors about the latest results of the cetacean–human communication project. He adjusts the microphone and begins to speak.

Scenario: Sanctuary

Dear investors, welcome to this meeting. In a short while, I'll report to you the results of our efforts of the last five years. I'm honoured to be here today to begin a process of re-thinking tourism and our relations with the pod of orcas with whom we have communicated for the past several years. We can now reasonably believe that collaboration between local tourism operators and this specific pod is possible and that such collaboration could include the development of tourism products that might benefit both us and them.

People in the conference room are visibly excited to learn more about this. After a short pause, Dr Brin continues:

We have identified some values in the orcas' culture that are very much in line with ours. This particularly concerns the ways in which individuals who might be in need of help are assisted by other individuals of the same group. Let me explain in greater detail. The orcas have shown interest and also a sort of admiration in our efforts to give elderly and children the assistance they need. They have also shown interest in our medicine, particularly veterinary medicine. Based on this, the orcas' proposal is to establish a sort of sanctuary along the northern Norwegian coast. This sanctuary would host orcas who might not be fit to live in the wild. The orcas have informed us that, at the moment, there are two calves and one old member of the pod who are in this condition. Their proposal is that we take care of these animals on behalf of the pod. The orcas have informed us that this possibility has long been discussed and that the interested individuals, including the calves' mothers, have agreed to it.

An investor asks:

So …we could establish a sanctuary and …sorry… I don't see how this will be lucrative for us … I mean, will the sanctuary be open to whale-watching boats? How will our situation, strictly from a business point of view, improve? For example, thinking about these three orcas you mentioned… what is the likelihood that tourists will manage to see them in such a vast area?

Dr Brin answers:

The orcas' proposal is interesting, but there are some practical challenges that should be considered. The answer I can give to you is that those individuals will need to be familiar with humans, accept us being close, sometimes very close, to them. This is because they might need some medical care. This can also be used to allow tourists to enjoy close encounters with the animals. The orcas have also clarified that the pod would visit those individuals living in the sanctuary: the hosted orcas will not be abandoned by their families but will be visited regularly.

This could be used to develop tourism products with a strong educational component that goes beyond the animals and the environment and encourages the tourists to reflect on care as the fundamental value of every society.

Scenario: Who Watches Whom?

Ladies and gentlemen, it is an honour for me to be here today to report the result of our project. In recent years, the orcas have received plenty of information about all human activities involving them. They have informed us that they had observed us for a long time and they were happy to have the opportunity to communicate with us. They have shown quite a lot of interest in the practice of keeping wild animals in captivity. At first, we had a hard time in explaining to them the logic behind keeping some orcas in captivity and training them to perform for humans.

The investors are listening attentively and wondering where this is going. Dr Brin clears his throat and says:

It seems that there is no agreement among the orcas about a possible future collaboration with us. Some orcas are not interested at all; others are more amenable. The latter group is composed mainly of young animals.

After a brief pause, Dr Brin continues:

The orcas willing to establish collaborative relations with us are very curious about humans. They propose that we build a sort of underwater human zoo where they can observe humans. They would like to have the possibility to observe us engaged in different activities, in particular dancing, eating and nursing our babies. They would prefer to have the possibility of observing the same people over long periods of time, so that they may become familiar with them and establish some sort of affective bond.

A woman raises her hand and speaks:

This does not sound right! I mean, the captivity part is just unacceptable... What kind of collaboration is this?!

More positively, another investor comments:

We should try to be constructive, maybe there are possibilities... I'm thinking about an underwater building including restaurants, shops, museums and hotel rooms. Tourists might spend their holidays there, and orcas could observe them while they were engaged in various activities. It sounds weird... but maybe feasible. And people could observe the orcas in return.

Scenario: Shattered Hopes

Dear investors, first of all let me thank you again for your interest and support. We are now in a position to present our conclusions about the possibility of establishing a collaborative relationship with the pod of orcas with whom we have been communicating over the last five years.

Dr Brin feels nervous and wonders how the investors will react to what he is about to tell them. He takes a sip from the glass of water in front of him, adjusts the microphone again and says:

The orcas do not want to collaborate with us. They do not see how they could benefit from such a collaboration, either in general or in relation to tourism.

Everybody in the conference room is silent. An investor raises his hand, and Dr Brin invites him to speak. The investor says:

Are you sure about your conclusions? Collaboration should be possible... I mean, orcas are clever, aren't they? I mean, we should try to find a common vision... and tourism might be a part of it...

Dr Brin answers:

Yes, we are sure about our conclusions. And, yes, orcas are indeed very clever animals. We spent quite a lot of time trying to explore possibilities and common grounds for collaboration. The orcas told us that they have been observing us, and don't want to have anything to do with us.

Another investor asks:

What do you mean that they don't want to have anything to do with us? Aren't they at least interested in humanity, our culture?

Dr Brin clarifies this point for the audience:

Yes, this is exactly the point. They are not interested in humanity. They consider our culture very primitive. They view us, humans, as extremely unreliable beings.

Discussion of the 2055 Scenarios

The scenarios described in the above sections offer some important points of reflection about the wildlife tourism of today and tomorrow. The future described in the first 2055 scenario, Sanctuary, is centred on the concept of mutuality and highlights the value of caring relations. As social animals and mammals, humans and orcas might have common ground on which to establish meaningful and fruitful relationships that might lead to mutual benefits. The orcas of this future express their desire for human interference in their lives when such interference is intended to help them. They view human veterinary medicine as a valuable resource from which they might benefit and request human help.

Dr Brin and the investors from the Sanctuary scenario seem to be optimistic. In particular, Dr Brin highlights the possibility that this form of tourism might truly be educational and transformative. This gives rise to the question of whether the investors from the tourism sector and the scientists will manage to establish such a sanctuary. This might present several challenges, chief among which will be the presence of other human activities in the area, in particular oil exploration and drilling. This highlights the importance of considering the future of the tourism sector, including its interrelation with other sectors. In particular, non-captive wildlife tourism requires the preservation of wild habitats and thus may conflict with other human activities. Therefore, considerations in terms of how to communicate the value of wildlife tourism and highlight its lucrative potential in comparison with other sectors, both in the short and long term, are not to be underestimated.

In the 2055 scenario Who Watches Whom?, some orcas propose human captivity. This provokes a negative reaction from the audience, a reaction that is quite easily understood and which can be related to critics' current concerns about captive settings for animals. In this scenario, an investor uses his creativity to think about a compromise, a solution that might respond to the orcas', the operators' and the tourists' desires. This is also a form of mutuality, albeit different from that seen in the previous scenario. In sacrificing some of their privacy, humans might be able to satisfy the orcas' curiosity as well as their own. Would the tourists be willing to do so? We may conjecture that some tourists might be willing to pay for an experience of this nature. The Who Watches Whom? scenario invites us to consider different types of wildlife tourists. In the same vein, it also demonstrates that, within that specific pod of orcas, different concerns and opinions are at play. Only some orcas are interested in observing the humans. The orcas, just as the tourists, may be an inhomogeneous group in relation to their needs, desires and perspectives. This indicates the necessity of including considerations at the individual level when planning activities that involve animals.

The third 2055 scenario, entitled Shattered Hopes, is the most provocative. It suggests that we might overestimate ourselves, and the invitation to look inward is clear. What, in our opinion, distinguishes us from other animals? Based on which specific ability do we tend to consider humanity superior to other forms of life? The Shattered Hopes scenario suggests that the animals might not perceive any value either in encounters with humans or in humanity as a whole. If this is the case, why should we impose our unwelcome presence on the animals? How can we justify such interference? Our best option may be to considerably limit wild animal–human encounters, both in general and in the context of tourism.

Conclusions

The idea of interspecies communication opens up a universe of future possibilities. This chapter has focused on orca-based tourism along the Northern Norwegian coasts and on three alternative scenarios envisioned as occurring in 2055. This chapter concludes by identifying the most important lessons that can be learned from these imagined futures. These concern the importance of (1) highlighting and communicating the potential lucrative advantages of the desired situations, (2) considering and respecting each animal's individuality and (3) reflecting on the possibility and opportunity to refrain from further developing tourism.

One key lesson that may be learned from the first scenario is that the lucrative aspect of future developments of the wildlife tourism sector should not be underestimated as the sector competes with other sectors in relation to various resources, for example, space. Futures that might be more desirable from an ethical standpoint run the risk of never becoming reality if their potential in terms of profit is not particularly high and not effectively communicated to the relevant actors, such as investors, other economic sectors and the government.

An important lesson that emerged from the second scenario is that considering that human experiences are highly subjective, so too might animals' experiences be. Wild animals can be differently affected by tourism activities, depending on their species and also on their individual characteristics. Consequently, animals should be given the greatest degree of freedom to choose whether and how they interact with tourists, both in captivity and in the wild, where the animals' individual characteristics are usually unknown to the tourism operators.

The key lesson that can be learned from the third scenario concerns wildlife tourism's fundamental basis as an invasive form of human activity in the animal world. It might be opportune to consider whether tourism activities should be strictly limited or completely prohibited, as this might be the animals' desire.

The final reflection, which concludes this chapter, concerns the opportunity that we might avail of, before stepping into the animal world, to reflect on what the animals might think about us. Wildlife tourism encounters are encounters between subjects with different perspectives; Derrida's cat's gaze reminds us of this. Animals might never acquire the competence to communicate their opinions on humans and tourism to us, and we might never develop the capacities required to understand the animals' perspective, but where does this leave our sensibility and sense of responsibility? Maybe it is not the lack of interspecies communication that prevents us from hearing the animals' voices. Relying on our sensibility and sense of responsibility, we might actually hear their voice, understand their perspective and, maybe, choose to act accordingly.

Acknowledgement

I gratefully thank Dr Heike I. Vester for her inspiring and responsible engagement in understanding and protecting whales and dolphins in the challenging context of Northern Norway.

References

Baker, R. (1974) The Great Whales' Mistake. *New York Times*, 23 June 23, p. 6. See https://www.nytimes.com/1974/06/23/archives/the-great-whales-mistake-sunday-observer.html (accessed 15 August 2020).

Ballengee-Morris, C. (2002) Cultures for sale: Perspectives on colonialism and self-determination and the relationship to authenticity and tourism. *Studies in Art Education* 43 (3), 232–245.

Bertella, G. (2017) Factors of peripherality: Whale watching in Northern Norway. In Y.-S. Lee, D. Weaver and N. Prebensen (eds) *Arctic Tourism Experiences Production, Consumption and Sustainability* (pp. 130–139). Oxford: CABI.

Bertella, G. and Vester, H.I. (2015) Whale watching in Norway caught between more traditional hunting canons and the lucrative promise of seismic airguns. *Tourism in Marine Environments* 11 (1), 73–77.

Bertella, G., Fumagalli, M. and Williams-Grey, V. (2019) Wildlife tourism through the co-creation lens. *Tourism Recreation Research* 44 (3), 300–310.

Brakes, P., Dall, S.R., Aplin, L.M., Bearhop, S., Carroll, E.L., Ciucci, P., Fishlock, V., Ford, J.K.B., Garland, E.C., Keith, S.A., McGregor, P.K., Mesnick, S.L., Noad, M.J., di Sciara, G.N., Robbins, M.M., Simmonds, M.P., Spina, F., Thornton, A., Wade, P.R., Whiting, M.J., Williams, J., Rendell, L., Whitehead, H., Whiten, A. and Rutz, C. (2019) Animal cultures matter for conservation. *Science* 363 (6431), 1032–1034.

Bramwell, B. and Lane, B. (eds) (2000) *Tourism Collaboration and Partnerships: Politics, Practice and Sustainability*. Clevedon: Channel View Publications.

Campos, A.C., Mendes, J., Oom do Valle, P. and Scott, N. (2017) Co-creating animal-based tourist experiences: Attention, involvement and memorability. *Tourism Management* 63, 110–114.

Carpenter, E. (1956) The timeless present in the mythology of the Aivilik Eskimos. *Anthropologica* 1–4.

Cater, C. (2010) Any closer and you'd be lunch! Interspecies interactions as nature tourism at marine aquaria. *Journal of Ecotourism* 9 (2), 133–148.

Convention on Migratory Species (2017) Recreational In-Water Interaction with Aquatic Mammals. Report prepared by the Aquatic Mammals Working Group of the Scientific Council. UNEP/CMS/COP12/Inf.13

Curtin, S. (2006) Swimming with dolphins: A phenomenological exploration of tourist recollections. *International Journal of Tourism Research* 8 (4), 301–315.

Deecke, V.B., Ford, J.K. and Spong, P. (2000) Dialect change in resident killer whales: Implications for vocal learning and cultural transmission. *Animal Behaviour* 60 (5), 629–638.

Department of Fisheries (2018) Stor risiko for ulykker med hvalturisme [High risk of accidents in whale watching tourism]. See https://www.fiskeridir.no/Yrkesfiske/Nyheter/2018/1118/Stor-risiko-for-ulykker-med-hvalturisme (accessed 15 August 2020).

De Waal, F. (2016) *Are We Smart Enough to Know How Smart Animals Are?* London: Norton.

Desmond, J. (1999) *Staging Tourism: Bodies on Display from Waikiki to Sea World*. Chicago, IL: University of Chicago Press.

Gaard, G. (1993) *Ecofeminism: Women, Animals, Nature*. Philadelphia, PA: Temple University Press.

Gibson, C. (2010) Geographies of tourism: (Un)ethical encounters. *Progress in Human Geography* 34 (4), 521–527.

Gruen, L. (2015) *Entangled Empathy: An Alternative Ethic for Our Relationships with Animals*. Brooklyn, NY: Lantern Books.

Higham, J., Bejder, L. and Williams, R. (2014) *Whale-Watching: Sustainable Tourism and Ecological Management*. Cambridge: Cambridge University Press.

Leslie, H., Banks, G., Prinsen, G., Scheyvens, R. and Stewart-Withers, R. (2018) Complexities of development management in the 2020s: Aligning values, skills and competencies in development studies. *Asia-Pacific Viewpoint* 59 (2), 235–245.

Lück, M. and Jiang, Y. (2007) Keiko, Shamu, and friends: Educating visitors to marine parks and aquaria? *Journal of Ecotourism* 6 (2), 127–138.

Mann, J., Connor, R.C., Tyack, P.L. and Whitehead, H. (2000) *Cetacean Societies: Field Studies of Dolphins and Whales*. Chicago, IL: University of Chicago Press.

Marino, L., Connor, R.C., Fordyce, R.E., Herman, L.M., Hof, P.R., Lefebvre L., Lusseau, D., McCowan, B., Nimchinsky, E.A., Pack, A.A., Rendell, L., Reidenberg, J.S., Reiss, D., Uhen, M.D., Van der Gucht, E. and Whitehead, H. (2007) Cetaceans have complex brains for complex cognition. *PLoS Biol.* 5 (5): e139. https://doi.org/10.1371/journal.pbio.0050139

Montford, K.S. (2016) Dehumanized denizens, displayed animals: Prison tourism and the discourse of the zoo. *PhiloSOPHIA* 6 (1), 73–91.

Parry, D.C. and Johnson, C.W. (2007) Contextualizing leisure research to encompass complexity in lived leisure experience: The need for creative analytic practice. *Leisure Sciences* 29 (2), 119–130.

Parsons, E.C.M. and Brown, D. (2018) Recent advances in whale watching research: 2016–2017. *Tourism in Marine Environments* 13 (1), 41–51.

Parsons, E.C.M., Fortuna, C.M., Ritter, F., Rose, N.A., Simmonds, M.P., Weinrich, M., Williams, R. and Panigada, S. (2006) Glossary of whale watching terms. *Journal of Cetacean Research and Management* 8 (Supplement), 249–251.

PETA (2019) SeaWorldofHurt. See https://www.seaworldofhurt.com/features/ (accessed 15 August 2020).

Phi, G.T. and Dredge, D. (2019) Collaborative tourism-making: An interdisciplinary review of co-creation and a future research agenda. *Tourism Recreation Research* 44 (3), 284–299.

Pryor, K.W. (1990) Non-acoustic communication in small cetaceans: Glance, touch, position, gesture, and bubbles. In T.J.A. and R.A. Kastelein (eds) *Sensory Abilities of Cetaceans* (pp. 537–544). Boston, MA: Springer.

Rendell, L. and Whitehead, H. (2001) Cetacean culture: Still afloat after the first naval engagement of the culture wars. *Behavioral and Brain Sciences* 24, 360–373.

Richardson, L. and St. Pierre, E.A. (2005) Writing: A method of inquiry. In N.K. Denzin and Y.S. Lincoln (eds) *The Sage Handbook of Qualitative Inquiry* (pp. 959–978). Thousand Oaks, CA: Sage.

Scheyvens, R., Banks, G. and Hughes, E. (2016) The private sector and the SDGs: The need to move beyond 'business-as-usual'. *Sustainable Development* 24 (6), 371–382.

Whitehead, H., Rendell, L., Osborns, R.W. and Würsig, B. (2008) Culture and conservation of non-humans with reference to whales and dolphins: Review and new directions. In S. Armstrong and G. Botzler (eds) *The Animal Ethics Reader* (pp. 181–192). Oxford: Routledge.

Williams, M. (2006) *Greek Myths*. London: Walker Books Limited.

Wiener, C. (2013) Friendly or dangerous waters? Understanding dolphin swim tourism encounters. *Annals of Leisure Research* 16 (1), 55–71.

Yeoman, I. and Mars, M. (2012) Robots, men and sex tourism. *Futures* 44 (4), 365–371.

Part 3
Technology Advancements

9 Safeguarding Sustainable Futures for Marine Wildlife Tourism through Collaboration and Innovation: The Utopia of Whale-Watching

Hindertje Hoarau-Heemstra and
Anne-Mette Hjalager

Commercial tours that enable tourists to observe, swim with and/or listen to any whale, dolphin or porpoise species in their natural habitat, is considered whale watching tourism (O'Connor *et al.*, 2009). Whale watching has expanded into a massive and highly profitable industry in many coastal communities all over the world and is an integral aspect of local economies (Lambert *et al.*, 2010). Whale watching can provide many socioeconomic benefits for communities and could potentially aid conservation as well. Thus, whale watching is seen by many environmental and animal welfare groups as a sustainable alternative to commercial whaling (Parsons, 2012). The public's desire to see and interact with whales and dolphins comes with a responsibility to protect them (New *et al.*, 2015). However, rapid development in many areas has challenged the industry's long-term sustainability (Higham *et al.*, 2014; Garrod & Fennell, 2004). In order for whale watching to progress to a sustainable pathway, the most urgent priority is to afford cetaceans meaningful protection from the causes of significant anthropogenic impact (Higham *et al.*, 2014).

A number of factors influence the impact of whale watching, including the types of vessels used, and there are many direct and indirect impacts on the target species. In 2006, the International Whaling Commission (ICW) stated that there is new and compelling evidence that the fitness of individual odontocetes (toothed whales) repeatedly exposed to whale watching vessel traffic can be compromised and that this can lead to effects on the population level. Marine mammals have been reported to change behaviours such as feeding or resting, and documentation exists that they have temporarily or permanently abandoned habitats (Bejder *et al.*, 2006; Carrera *et al.*, 2008; Lusseau, 2005). Research also shows that boat-related sounds can drown out or mask cetacean vocalisations, and animals will then either be unable to communicate or have to increase the volume of their vocalisations (Guerra *et al.*, 2014). However, the animal response depends on the type of vessel and its behavior (Erbe, 2002). Pollution in the form of exhaust emissions from whale watching vessels can also affect the exposed individuals (Lachmuth *et al.*, 2011).

Overall, the populations and ecosystems might be stable but there are ethical issues with disturbing individual animals. Sheppard and Fennell (2019) show that tourism policy is increasingly considering a broader and deeper range of impacts, including concern for individual animals. This is a step in a more ethical direction where tourism policy and behaviour will eventually evolve to a level where human and animal welfare and rights are considered in tandem.

Taking into consideration the aforementioned animal welfare and rights discussions, in addition to climate change prospects, marine biodiversity challenges and potential politically and economically instigated restrictions on travel, it can be asked whether whale watching tourism is likely to exist unchanged in the future? As marine tourism is a noteworthy economic factor in many peripheral coastal communities, it is critical to assess the particularities of any measures that adapt the activity or the resource in order to ensure more sustainable operations. While most studies address the habitats and welfare of the whales, the tourist experience and provided services and governance framework (Higham *et al.*, 2014: chapter 25), there is less knowledge about the economic and technological driving factors and future framework conditions for more sustainable whale watching (Higham *et al.*, 2014).

Approach to Exploring the Future of Whale Watching

Based on the case study of the Icelandic whale watching company Friends of Whales (FoW) (presented under a pseudonym), this chapter reflects on the role that a tourism wildlife business can play in terms of taking the initiative to discover technological eco-innovations that may change the business of wildlife watching in the future. In 2010, 2011 and 2017, one of the authors of this chapter carried out a qualitative study on innovation in whale watching where FoW was a participating company. Further discussions of the company and a future scenario are based on the data gathered in this study.

FoW specialises in delivering whale watching tours across the North Atlantic. It operates from Husavik (Northern Iceland), which is one of the most popular destinations in Northern Europe to see cetaceans in their natural habitat. FoW was the first mover in a technological co-industry eco-innovation that provides existing whale watching boats with electric propulsion instead of being powered by diesel engines. The development included potential prototypes for solutions in critically fragile environments; therefore, the company and its collaborators were frontrunners in the search for novel sustainable innovation practices in tourism.

The exemplary value of the collaborative processes in this particular context provides tools for academic and applied thinking about future innovation systems in tourism and, for that reason, the FoW case is informative in this context. In the remainder of the chapter, we discuss innovation and collaboration in the context of wildlife tourism. We further present a utopian future scenario of whale watching with FoW in the year 2050. Our story is utopian because we present wildlife tourism as an activity that is better than what we find today, meaning that death and suffering are absent and that the relationship between humans and wildlife has changed with a focus on equality. As Victor Hugo said in *Les Misérables*: 'There is nothing like a dream to create the future. Utopia today, flesh and blood tomorrow'. Therefore, we consider our chapter to be predictive in nature as our arguments are based on strong, explained and truthful evidence. Understanding and framing the future of sustainable whale watching by looking at today's innovations are important as the future is the only thing one can change, influence or create (Yeoman & Postma, 2014). Following the presentation of our utopian future scenario, the relevant drivers of change are discussed in detail.

Innovation and Collaboration

Global tourism can be seen as a highly complex and open system with a multitude of actors who interact at cross-cutting levels (Cornelissen, 2005; Hall, 2004). The whale watching industry is part of this system and interacts with a wide range of external forces, both directly and indirectly related to tourism, in a dynamic manner (Higham *et al.*, 2014).

As with other forms of tourism, the future of wildlife tourism, whale watching included, depends on the innovations and changes introduced today. To consolidate knowledge and gain inspiration, future studies, scenario building and prognoses are therefore known elements in innovation management processes (Rodriguez *et al.*, 2014; Yeoman, 2012). Innovation refers to the process of bringing any new, problem-solving idea into use (Hall & Williams, 2008). Based on Schumpeter's theory of entrepreneurship and economic development from 1934, the innovation may be assigned as a product innovation, an organisational innovation (a new collaborative structure) or a process innovation (a new kind of service delivery) (Rønningen & Lien, 2014). The ways that innovations are developed in the tourism industry depend on four distinctive features: close interaction between consumers and producers of tourism products; information intensity; the importance of the human factor; and the critical role of organisational factors (Hall, 2008).

Tourism businesses seem to be less innovative than enterprises in the manufacturing, financial and other service sectors (Camisón & Monfort-Mir, 2012; Hjalager, 2010). Also, tourism enterprises are not recognised as technological innovators to any critical extent, as they focus on services and tourism encounters, and they often work unsystematically (Nordin & Hjalager, 2017). Paradoxically, there is a critical need for innovation in tourism where newness is a competitive advantage and sustainability challenges need addressing.

Divisekera and Nguyen (2018) found that collaboration is the most important determinant for innovation in the tourism sector. Firms taking part in collaborations are more likely to introduce innovations than firms that do not collaborate (Buijtendijk *et al.*, 2018; Makkonen *et al.*, 2018). Hence, being a part of a network can offer clear benefits to small and medium tourism enterprises (Gomezelj, 2016; Inkpen & Tsang, 2005) in helping them decode and appropriate flows of knowledge for innovation (Hoarau & Kline, 2014). The tourism innovation literature (Kuščer *et al.*, 2017) shows that collaboration for innovation mainly takes place with other tourism enterprises, through destination management bodies and constellations of customers. Tourism businesses seldom grab opportunities for collaboration with suppliers and advanced technology actors and these types of collaborations are also less understood (Beritelli, 2011; Carlsen & Edwards, 2008; Hjalager, 2015).

Supported by the results of a systematic literature review on environmentally sustainable product innovation conducted by De Medeiros *et al.* (2014), alignment with suppliers seems an important strategy to overcome the deficiencies of technological issues that hamper the eco-innovation activities of tourism enterprises. However, collaboration can complicate the management and organisation of innovation as relevant knowledge is widely disseminated within and outside of organisations. Knowledge flows that support innovation are particularly governed by social norms such as trust between partners (Zach & Hill, 2017). Trustful cooperative behaviour creates a basis for knowledge transfer and learning between tourism organisations across borders (Makkonen *et al.*, 2018). Zack and Hill (2017) found that a tourism firm's choice of innovation partners is primarily associated with current collaboration patterns, shared knowledge and relational trust. Hence, like-minded people find one another and a common cause, and their practice-based knowledge sharing contributes to the development of tourism (Hoarau & Kline, 2014).

Visioning a Utopian Scenario for Whale Watching in Iceland in the Year 2050

Zoë, a 37-year-old European woman, boards a two-masted schooner in the modern facilities of the Husavik harbour in Northern Iceland. It is summer, and she is wearing light clothes suitable for the pleasant

Arctic summer temperatures. She thinks that 25 degrees at sea feels so much nicer than the 45 degrees back home in London. She has been looking forward to this moment for a very long time. She has always loved whales and dolphins and is very engaged in their well-being. Worldwide, cetaceans are doing better now since the massive clean-up of the world's oceans in the last decades, but climate change keeps reducing their numbers. Although they are now highly protected worldwide, Zoë just learned that whales were hunted and eaten in Iceland until 2025. She can hardly believe that previous generations did not understand the intrinsic value of animals and their rights of protection. Other than a few extremists, most people now follow a vegan lifestyle. Some animal species, e.g. cetaceans and primates, obtained animal personhood in 2030. Several non-governmental organisations (NGOs) are representing these animals' rights based on extensive scientific data and improved interspecies communication.

Enhanced animal rights were a hard blow for FoW, the company that offers trips to meet what remains of the Icelandic whales. When starting operations in the early 1990s, the company chased whales using traditional oak fishing boats powered by diesel engines. There were several companies in Husavik at that time, and during summer at least 20 boats would be in the water from early morning to midnight. They were all looking for the blue whales, dolphins, porpoises, mink whales and humpback whales that came to feed in the bay. The diesel engines polluted the water and the air of the bay and made a lot of noise. Zoë has just learned that scientific studies since the beginning of the century demonstrated that the continuous presence of diesel engines was detrimental to the health of the whales in the bay. They suffered from the pollution of the air and water and the noise inhibited their communication. In addition, the presence of boats interfered with their feeding and resting behaviour. All these issues improved following the introduction of regulations on whale watching under global rules in 2030. The collaboration and lobbying of animal rights groups, whale watching companies and governments established a global governance work group to produce a whale protection plan. FoW played a central role as representative of the European whale watching businesses. The company always maintained an extensive network of like-minded sectors consisting of NGOs, governmental organisations and responsible businesses.

In 2050, whale watching is no longer a mass tourism product. Now, companies have to earn their revenue with a limited amount of people and trips per day. Every area has a physical as well as a time sanctuary. As a result, there are long waiting lists and whale watching has become a very exclusive and expensive attraction.

Zoë saved for the trip over five years and was on a waiting list for three years. On board the schooner, she thinks the vessel has come straight from a history book as it has been in whale watching service for nearly 40 years now. The sails are used when there is enough wind, taking the passengers back to a time centuries ago. The engine is powered by the latest technology. Propellers are no longer used as these can hurt the whales and interfere with their sonar systems. An automatic pilot identifies where the whales are located, what they are doing and how near they can be approached. The speed at which to approach whales is programmed not to disturb the animals. The tourists on board follow the action with special glasses that show holograms of underwater drones that get close to the whales. The tourists can admire the whales on the water's surface and also observe the animals underwater. They can listen to the sounds the animals make and, recently, understand what they say. The code of whale language has just been cracked by scientists who are recording their sounds during whale watching trips in the bay. The whales talk about where to find food, warn each other about the presence of humans and exchange the latest gossip of the pod. Zoë admires their beauty and intelligence and thinks about the old days of whale watching when these creatures were not understood and were bothered by humans wanting to watch them, and worse. Unthinkable for a generation that has grown up recognising the personhood of cetaceans.

On the return trip to the harbour, Zoë turns on the information feed in her glasses and learns about FoW's evolution from mass tourism on oak diesel boats, to highly exclusive and high technology tourism in 2050. The company managed to survive for almost 70 years thanks to collaboration, learning and innovation. The proprietors were well-prepared for the new rules because they had always considered animal welfare and environmental impacts. Many other whale watching companies worldwide did not survive because they struggled to adapt once the new rules were in place and enforced by the android coastal police that is able to overrule the captain and send the boat back to the harbour in cases of non-compliance with the rules. Now the cetaceans of the world have a chance to come back from their second near-extinction experience.

Signals and Signposts as Drivers of Change

The scenario presents a range of fundamental changes to whale watching today as a consequence of increasing environmental awareness, worsening climate crises, equality between species that give cetaceans rights and the use of technology for finding, experiencing and interacting with cetaceans. This has led to new rules and regulations for the tourism industry. Seeking developments that signal change, the authors allow for new schemes of interpretation.

In pursuit of evidence of futures, signposts and signals may provide patterns that suggest the directions ahead (Rosenau, 1995). In order to discuss signals and signposts relevant to our scenario, we adopt the methodology of Robertson and Yeoman (2014). The approach raises a series of questions that analyse signposts and signals for the future of whale watching. Yeoman *et al.* (2012) argue that signposts are the truthfulness, occurrence and evidence of current scenarios, while signals are indicators of what the future could be. We use the term 'signposts' to indicate events and developments associated with a particular scenario, and 'signals' to indicate the characteristics of whale watching tourism in 2050.

In our study, the signposts relevant to change in the whale watching sector are: the diffusion of concerns about environmental ethics, animal rights and more ethical practices in the tourism industry; the innovation of electric engines in whale watching vessels; the use of immersive technology; the identification of partners in the innovation process; and learning from innovation. These signposts indicate that emergent collaboration across industries, new forms of technology, new demands from customers and pressure from the climate change lobby will gather along the trajectory of the whale watching industry. The relevant signals will establish and develop connections in the whale watching industry. Further, such signals in this scenario highlight the gradual change in the consumption of nature-based experiences. New questions increasingly address how eco-innovations are spread from the first movers to less-developed nations and companies so as to obtain further benefits.

Change Drivers in Whale Watching

Ethics on the rise

As the scenario suggests, the rapid increase in the number of whale watching tourists during the first two decades of the 21st century was not likely to continue as this growth was based on unsustainable business operations. Higham *et al.* (2014: 376) argue that in order for whale watching to be sustainable, there has to be a recognition of the intrinsic worth of marine mammals that should be offered protection for their own sake, and left to their own devices for significant periods of their lives, free from human demands.

Due to near extinction following pollution and climate change, we propose that there are signals that environmental ethics and animal rights will evolve. Innovation can improve the interaction between humans and cetaceans and allow for more ethical whale watching. The technology developed by FoW and its partners created a new type of tourism product: silent and low-emission whale watching. In this case, the innovation and implementation of new technologies improved the experience for all parties involved, allowing tourists to experience wildlife with minimal impact. This development can be referred to as the maturing of the tourism industry to the point where it is now willing to discuss both the negative and the positive impacts of tourism. This is in line with the considerations of Sheppard and Fennell (2019) who argue that the tourism industry is developing moral consciousness, particularly with regard to the ethical issues associated with the use of animals as part of the tourism experience.

Eco-innovations: Electric propulsion and immersive technology

One central signpost is the innovation of electric engines in whale watching vessels. In Iceland, the transition from diesel to electric started with an innovation project that was financially supported by the Icelandic government and managed by FoW. Since the 1990s, FoW has offered whale watching tours through Skjálfandi Bay (Northern Iceland), in Tromsø (Norway) as well as multiple day-expedition cruises to Greenland. High on FoW's strategic agenda has been alternative transportation fuels. Price volatility and issues related to scarcity and the environmental impacts of fossil fuels were pertinent concerns for the company. However, few evaluations had been done on alternative fuels for marine vessels, and for FoW the market did not provide solutions to transform whale watching boats from diesel-fuelled to sustainable vessels. Therefore, in 2012, FoW initiated an innovation project with the aim of developing a propulsion system that would improve the financial performance, the environmental influence, the impacts on wildlife as well as the touristic experience. The goal of the project was to develop a technology that would it make possible to store electric power either from shore or generated by the propeller of sailing boats on board vessels to gain silent and clean propulsion. For a ship with sails, regeneration adds the benefit of capturing energy generated by the sails during good wind conditions for recharging batteries.

Other potential technologies that have attracted attention in recent times are immersive technologies such as augmented reality (AR), mixed reality (MR) and virtual reality (VR), which are predicted to have a profound influence on the future of the tourism industry (Guttentag, 2010). Immersive technology refers to technology that attempts to emulate a physical world through a digital or simulated world, thereby creating a sense of immersion. These technologies differ in terms of how much of the 'real' world is represented. AR overlays the real world with 2D or 3D computer-generated information; VR is a virtual computer-simulated world; and MR combines real and virtual objects with which the user can interact (Beck *et al.*, 2019).

Wildlife tourism and whale watching in particular are challenged because nature cannot be controlled. Animals are observed in their natural habitat which can disappoint the customers when they do not have the expected interaction. Whale watching enterprises have sought innovative solutions since the beginning of whale watching (Hoarau & Kline, 2014) and immersive technologies might be the solution. Customers can learn, observe and interact with wildlife via AR, MR and VR by wearing glasses, looking at holograms or being part of a virtual world. In order to develop these kinds of programmes and products, the tourism industry will have to work closely with programmers, artists and engineers.

Identification of partners

The third signpost identified in this chapter is the building and maintenance of a network for innovation that brings partners from different industries together. For that purpose, FoW approached an environmental NGO and Icelandic New Energy (a public–private partnership) for assistance with identifying partners outside of their direct network. FoW highlighted the importance that all parties involved exhibited an understanding of the environmental and technical challenges of transforming a diesel-powered whale watching vessel into an electric ship. Collaboration between individuals proved to be essential, particularly when technology suppliers with different specialties were involved.

In addition to the importance of being equally enthusiastic and willing to try new things, it was considered critical that partners shared values and had a mutual understanding. For example, the partners had identical environmental considerations and motivations. They identified opportunities to sell their knowledge, to enter new markets or to contribute to a project that they viewed as their corporate social responsibility. The project clearly offered marketing advantages to all partners. Seeing and seizing opportunities are considered to be critical aspects of innovation processes.

Learning and collaboration for innovation

In order to achieve sustainability outcomes in the whale watching tourism industry, Higham *et al.* (2014) call for an all-inclusive, multi-stakeholder approach. This is also necessary for the development of eco-innovations that make whale watching operations more sustainable. Social capital is accumulated through interactions: interactions between individuals, between individuals and groups, and between groups of groups. Learning can (and frequently does) occur when individuals and groups interact (Kilpatrick *et al.*, 1999). Lundvall (1992) proposes that change is a cumulative process which builds on existing knowledge and practices through interactive learning. Hence, learning, building relationships and social capital can be seen as an investment in future development.

Learning was an essential element of the current innovation project discussed in this chapter because the technology as well as the context was new to the collaborating partners. Electric batteries and improved propulsion systems existed for the marine sector at the time, and to have a sailboat regenerate its own electric energy was revolutionary. The owners and employees of FoW understood how to sail boats, how to find whales and how to entertain tourists in the process, but had no technical background. The technology developers had experience with batteries and lorries, but little experience with sailing boats full of tourists in the Arctic. Moving away from their comfort zone and becoming involved in distinctive technological innovations required engagement in co-creative learning processes. For example, during the adjustments and tests of the prototype, technology developers were present on-board ships, necessitating the sharing of practice-based tacit knowledge with tourism actors. Exercising a sense of urgency, the project team was more resilient to setbacks, especially once they had established a pool of tacit and explicit collective knowledge on which they could rely. The uniqueness and originality of the innovative project sparked enthusiasm. Learning together and sharing values has built social capital and trust in this example of the whale watching innovation network and this can be used as a resource in subsequent innovation processes. We see innovation and change as an upward spiral where more interaction and learning leads to more innovation and so on. We therefore postulate that the pro-active, sustainable and ethical wildlife tourism enterprises of today will develop further in this direction based on their existing social capital. Social capital in the form of networks, norms and trust facilitates cooperation for innovation. Hence, future innovation is based on an evolutionary process for which the building blocks of knowledge and social capital are developed today.

Overcoming barriers

The fifth signpost in the utopian scenario is the ability to overcome barriers or to resist change. FoW is a profitable company that is able and willing to undertake risky investments. Consequently, the innovation proved to be more radical, and it changed the whole design of the boat and the experience for which the boat is used.

The cross-industry alliance proved to be very beneficial, but not without challenges, including the significant physical distance between those engaged in learning processes. Remote locations raise communication barriers, at least they do at the beginning of the 21st century. New communication technology might facilitate long-distance communication even more, so that social capital can be built between actors in different locations. This might also play a role in the dissemination of innovation between developed and developing countries.

Conclusion

The utopian scenario presented in this chapter demonstrates a successful eco-innovation process, and provides evidence that high-profile and highly profitable tourism companies can be first movers in a future orientation. It shows that tourism enterprises may, under the right circumstances – here competition, costs constraints and sustainable market demand – be motivated to engage in a long-term innovation project.

But will the future of whale watching be clean and have minimal impact on cetaceans? The diffusion of this eco-innovation might be a game changer for the tourism industry. The start of Icelandic whale watching was rather opportunistic with a focus on making as much money as possible. However, overcrowding, animal stress, pollution and increased focus on sustainability shifted the focus of entrepreneurs in wildlife tourism. The expectation is that tourists increasingly want to experience whales and dolphins in their natural habitat, and the North Atlantic will be one of the rare places on earth to do so because of the exploitation and climate stress that is threatening other aquatic and coastal ecosystems. Wildlife tourism businesses have a responsibility to future generations, wildlife and nature to ensure that further developments will be sustainable. Animal welfare and what is accepted as appropriate or not, in terms of human interactions with animals, is highly variable across cultures and times. That is why wildlife actors with social capital and networks are better able to understand changing trends and innovate accordingly.

In order to build social capital and be ready for wildlife tourism in the second half of the 21st century, businesses will need to extend their boundaries and develop collaboration with unknown stakeholders, such as knowledge intermediaries and technology suppliers. Tourism needs to develop a partnership approach for the future as the survival of the industry is at stake; in crises all parties survive by working together. Searching for external knowledge can be difficult due to the tacit, complex, competitive and indivisible nature of knowledge. FoW had access to limited information sources to scan and monitor its technological environments. However, through iterative steps whereby partners of partners were integrated into a viable project design, the stakeholders involved expanded their networks and knowledge bases.

Finally, it can be noted that in our utopian scenario many whale watching companies are out of business. These are the companies that did not have the resources to keep up with the aforementioned changes and were not able or willing to collaborate. This could have negative impacts on peripheral communities that rely on whale watching tourism. Thus, our scenario invites reflection on the various dimensions of sustainability and, eventually, what dimension and values to prioritise and how.

References

Beck, J., Rainoldi, M. and Egger, R. (2019) Virtual reality in tourism: A state-of-the-art review. *Tourism Review* 74 (3), 586–561.

Bejder, L., Samuels, A., Whitehead, H., Gales, N., Mann, J., Connor, R. and Kruetzen, M. (2006) Decline in relative abundance of bottlenose dolphins exposed to long-term disturbance. *Conservation Biology* 20 (6), 1791–1798.

Beritelli, P. (2011) Cooperation among prominent actors in a tourist destination. *Annals of Tourism Research* 38 (2), 607–629.

Buijtendijk, H., Blom, J., Vermeer, J. and Van der Duim, R. (2018) Eco-innovation for sustainable tourism transitions as a process of collaborative co-production. *Journal of Sustainable Tourism* 26 (7), 1222–1240.

Camisón, C. and Monfort-Mir, V.M. (2012) Measuring innovation in tourism from the Schumpeterian and the dynamic-capabilities perspectives. *Tourism Management* 33 (4), 776–789.

Carlsen, J. and Edwards, D. (2008) BEST EN case studies: Innovation for sustainable tourism. *Tourism and Hospitality Research* 8 (1), 44–55.

Carrera, M.L., Favaro, E.G.P. and Souto, A. (2008) The response of marine tucuxis (*Sotalia fluviatilis*) towards tourist boats involves avoidance behaviour and a reduction in foraging. *Animal Welfare* 17 (2), 117–123.

Cornelissen, S. (2005) *The Global Tourism System: Governance, Development and Lessons from South Africa.* Aldershot: Ashgate.

De Medeiros, J.F., Ribeiro, J.L.D. and Cortimiglia, M.N. (2014) Success factors for environmentally sustainable product innovation: A systematic literature review. *Journal of Cleaner Production* 65, 76–86.

Divisekera, S. and Nguyen, V.K. (2018) Determinants of innovation in tourism evidence from Australia. *Tourism Management* 67, 157–167.

Erbe, C. (2002) Underwater noise of whale-watching boats and potential effects on killer whales (*Orcinus orca*), based on an acoustic impact model. *Marine Mammal Science* 18 (2), 394–418.

Garrod, B. and Fennell, D.A. (2004) An analysis of whale-watching codes of conduct. *Annals of Tourism Research* 31 (2), 334–352.

Gomezelj, D.O. (2016) A systematic review of research on innovation in hospitality and tourism. *International Journal of Contemporary Hospitality Management* 28 (3), 516–558.

Guerra, M., Dawson, S., Brough, T. and Rayment, W. (2014) Effects of boats on the surface and acoustic behaviour of an endangered population of bottlenose dolphins. *Endangered Species Research* 24, 221–236.

Guttentag, D.A. (2010) Virtual reality: Applications and implications for tourism. *Tourism Management* 31 (5), 637–651.

Hall, C.M. (2004) *Tourism: Rethinking the Social Science of Mobility.* Harlow: Pearson.

Hall, C.M. (2008) Tourism firm innovation and sustainability. In D. Scott, M. Hall and S. Gössling (eds) *Sustainable Tourism Futures: Perspectives on Systems, Restructuring and Innovations* (pp. 282–298). New York: Routledge.

Hall, C.M. and Williams, A.M. (2008) *Tourism and Innovation.* New York: Routledge.

Hoarau, H. and Kline, C. (2014) Science and industry: Sharing knowledge for innovation. *Annals of Tourism Research* 46, 44–61.

Higham, J., Bejder, L. and Williams, R. (eds) (2014) *Whale-Watching: Sustainable Tourism and Ecological Management.* Cambridge: Cambridge University Press.

Hjalager, A.M. (2010) A review of innovation research in tourism. *Tourism Management* 31 (1), 1–12.

Hjalager, A.M. (2015) 100 innovations that transformed tourism. *Journal of Travel Research* 54 (1), 3–21.

Inkpen, A.C. and Tsang, E.W.K. (2005) Social capital, networks, and knowledge transfer. *Academy of Management Review* 30 (1), 146–165.

Kilpatrick, S., Bell, R. and Falk, I. (1999) The role of group learning in building social capital. *Journal of Vocational Education and Training* 51 (1), 129–144.

Kuščer, K., Mihalič, T. and Pechlaner, H. (2017) Innovation, sustainable tourism and environments in mountain destination development. *Journal of Sustainable Tourism* 25 (4), 489–504.

Lachmuth, C.L., Barrett-Lennard, L.G., Steyn, D.Q. and Milsom, W.K. (2011) Estimation of southern resident killer whale exposure to exhaust emissions from whale-watching vessels and potential adverse health effects and toxicity thresholds. *Marine Pollution Bulletin* 62 (4), 792–805.

Lambert, E., Hunter, C., Pierce, G.J. and MacLeod, C.D. (2010) Sustainable whale-watching tourism and climate change. *Journal of Sustainable Tourism* 18 (3), 409–427.

Lundvall, B.Å. (ed.) (1992) *National Systems of Innovation: Towards a Theory of Innovation and Interactive Learning*. London: Pinter Publishers.

Lusseau, D. (2005) Residency pattern of bottlenose dolphins *Tursiops* spp. in Milford Sound, New Zealand, is related to boat traffic. *Marine Ecology Progress Series* 295, 265–272.

Makkonen, T., Williams, A.M., Weidenfeld, A. and Kaisto, V. (2018) Cross-border knowledge transfer and innovation in the European neighbourhood. *Tourism Management* 68, 140–151.

New, L.F., Hall, A.J., Harcourt, R., Kaufman, G., Parsons, E.C.M., Pearson, H.C. and Schick, R.S. (2015) The modelling and assessment of whale-watching impacts. *Ocean & Coastal Management* 115, 10–16.

Nordin, S. and Hjalager, A.M. (2017) Doing, using, interacting: Towards a new understanding of tourism innovation processes. In A. Királóvá (ed.) *Driving Tourism through Creative Destinations and Activities* (pp. 165–180). Hershey, PA: IGI global.

O'Connor, S., Campbell, R., Cortez, H. and Knowles, T. (2009) Whale-Watching Worldwide: Tourism Numbers, Expenditures and Expanding Economic Benefits. A special report from Economists at Large, International Fund for Animal Welfare, Yarmouth, MA.

Parsons, E.C.M. (2012) The negative impacts of whale-watching. *Journal of Marine Biology* 2012, 1–9.

Robertson, M. and Yeoman, I. (2014) Signals and signposts of the future: Literary festival consumption in 2050. *Tourism Recreation Research* 39 (3), 321–342.

Rodriguez, I., Williams, A.M. and Hall, C.M. (2014) Tourism innovation policy: Implementation and outcomes. *Annals of Tourism Research* 49, 76–93.

Rønningen, M. and Lien, G. (2014) The importance of systemic features for innovation orientation in tourism firms. In G. Alsos, D. Eide and E.L. Madsen (eds) *Handbook of Research on Innovation in Tourism Industries* (pp. 27–55). Cheltenham: Edward Elgar Publishing.

Rosenau, J.N. (1995) Signals, signposts and symptoms: Interpreting change and anomalies in world politics. *European Journal of International Relations* 1 (1), 113–122.

Sheppard, V.A. and Fennell, D.A. (2019) Progress in tourism public sector policy: Toward an ethic for non-human animals. *Tourism Management* 73, 134–142.

Yeoman, I. (2012) *2050: Tomorrow's Tourism*. Bristol: Channel View Publications.

Yeoman, I. and Postma, A. (2014) Developing an ontological framework for tourism futures. *Tourism Recreation Research* 39 (3), 299–304.

Yeoman, I., Robertson, M. and Smith, K. (2012) A futurist's view on the future of events. In S. Page and J. Connell (eds) *The Routledge Handbook of Events* (pp. 507–525). London: Routledge.

Zach, F.J. and Hill, T. (2017) Network, knowledge and relationship impacts on innovation in tourism destinations. *Tourism Management* 62, 196–207.

10 Designing Future Wildlife Tourism Experiences: On Agency in Human–Sled Dog Encounters

Mikko Äijälä, Titta Jylkäs, Vésaal Rajab and Tytti Vuorikari

This chapter argues that the acknowledgement of agency contributes to more meaningful human–animal encounters in the context of wildlife tourism. We explore the agency of sled dogs in touristic mushing in order to create a space for re-examining the concept of wildlife tourism, which is often recognised as a subset of activities within nature-based tourism. These activities may be categorised according to impacts (consumptive/non-consumptive), venue (e.g. natural area, wildlife sanctuary or zoo/aquarium) or type of animals encountered (e.g. domesticated/non-domesticated). In addition to the commercialised visitor–wildlife encounters, the core of wildlife tourism comprises unintentional wildlife tourism experiences (Higginbottom, 2004). Due to the wide range of possible activities, venues and encounters, it has been challenging to develop a common definition for wildlife tourism (Skibbins, 2015). Nonetheless, wildlife tourism is founded upon the categorisation of certain animal species as 'wild' and as placed in 'wilderness' by human orderings (Buller, 2014).

Commonly, sled dogs are not associated with the category of wild, and mushing does not fall under the category of wildlife tourism. Mushing refers to a transport method or sport in which a dog or a team of dogs pulls a sled in snowy conditions, or a rig if there is no snow cover. Historically, sled dogs have represented the utopian ideal of trusted companions conquering the wilderness with 'a wolfish ancestry' (Onion, 2009: 154). Mushing as a touristic activity has a rather short history especially in Fennoscandia, but it has become one of the most popular touristic activities in some areas, such as Finnish Lapland (García-Rosell

& Äijälä, 2018). In the tourism context, mushing is a commercialised encounter between dogs, humans and the environment in order to create outdoor experiences. It is mainly based on encountering the charisma of a certain species of domesticated animal – specifically sled dogs – combined with the experience of an environment considered to be semi-wild or even tame (Bertella, 2016). Tourists' expectations of encountering almost wolf-looking dogs do not always meet reality, as individual dogs can differ from wolves by their looks. Despite their tameness, several of sled dogs' behaviour and social interactions are those of wolves, which are seldom reported in other dog breeds (Fiszdon & Czarkowska, 2008). Mushing might also involve encountering species of animals, such as elk, reindeer and fox, which are categorised as (semi-)wild. These encounters contribute to the overall outdoor experience (Curtin, 2009).

The experiences relate to individual human perception of sled dogs, and to the ways in which animals' environments are corporeally sensed and engaged with by the tourist through various activities and multiple senses (Ballantyne *et al.*, 2011). Therefore, wilderness and experiencing it through encountering certain species of animals is a human perception (Vannini & Vannini, 2016), which is not necessarily made up of the exotic but, instead, can take place through animal encounters in mundane settings (Curtin, 2010). In the future, tourists may be driven to seek out such experiences through an understanding of animals and their environment, and through engagement with non-human 'others' (Curtin, 2009). The experiences may also emerge through meaningful engagement with animals that are categorised as domesticated, but possess characteristics of wildness as a system that produces wilderness and wild things depending on the quality of the interaction from and among its components (Cookson, 2011). In the context of touristic mushing, sled dogs bridge the boundary between wild and artificial living, as tourists are caught between the security of readily provided resources and the temptation of wildness (Bertella, 2016).

Along with a desire to seek meaningful encounters with animals, there is a growing critical assertion that understanding the value of the human–animal encounters only through business-oriented terms neglects the ethics, agency and, at worst, the welfare of animals working in tourism (Fennell, 2012). Within the context of neglecting animal agency in wildlife tourism, we ask: How can the agency of sled dogs be evoked in a wildlife tourism context? How can the use of technology support the evoking of agency? How can design that acknowledges animal agency create new types of service value in the future? All the dogs (and humans) are individuals, who bring their own life history and experiences to the encounters and through their motile presence effect change on the institutional structures of tourism (Bertella, 2014; Notzke, 2019). Accordingly, we acknowledge that sled dogs working in tourism have agency (Buller, 2012; Koski & Bäcklund, 2017; Philo & Wilbert, 2000). Through recognition of sled-dog agency, we consider changes to critically examine the wildlife tourism concept in relation to the wildness of a particular environment.

We focus on the experiential aspect by adopting a speculative design approach (Dunne & Raby, 2013) to touristic mushing. Speculative design has been used as a tool for defining the design challenge and for ideating possible future scenarios. By exploring touristic mushing as a practice of interspecies relationship, we argue that sled dogs shape the wildness of a particular environment, which is an important factor of wildlife tourism. The study presents a narrative of potential futures and makes explanatory, but not truth, claims; this approach resonates with the critical use of science fiction as an ontological framework for tourism futures (Yeoman & Postma, 2014). The narrative suggests that the lack of mutual understanding may hinder more meaningful human–animal encounters. Rather than seeing sled dogs merely as objects of human desires to experience nature, however, our speculative concept illustrates what wildlife tourism could be like in the future if the animal with its practices is considered as a crucial agent. In doing so, the study contributes to envisioning future wildlife tourism as a space where collaborative encounters with animals are acknowledged (Lulka, 2004; Picken, 2018).

Experiencing the Environment Through Human–Sled Dog Encounters

Animal charisma is an essential asset in wildlife tourism, thus animals, such as polar bears, big cats and orcas, are appealing for tourists (Skibbins, 2015). For Lorimer (2007: 915), non-human charisma is the 'distinguishing properties of a non-human entity or process that determine its perception by humans and its subsequent evaluation'. Non-human charisma is partly a result of human perception, which means that the appeal of particular animals for particular tourists is based not only on the animals' natural characteristics, but also on the subjective expectations of a tourist. In addition to animal charisma, the environment, objects such as housing systems for animals and subjects such as the guiding musher all play a part in making a memorable experience (Bertella, 2014). However mundane an encounter with wildlife in someone's backyard or in a remote destination might be, it can contribute to an individual's experience of 'the wild' (Curtin, 2009).

Memorable experiences can be produced through multiple senses such as sight, sound, smell and touch (Ballantyne *et al.*, 2011). When animal charisma is accompanied by perceived natural settings and factors, such as an emotional connection, the human–animal encounter may lead to a memorable experience (Hughes, 2013). When encounters are directly 'in nature itself', such connections may be amplified (Bentrupperbäumer, 2005: 88). The attractiveness of mushing is partly based on representations of exotic environments, which are produced in tourism marketing that commodifies nature as wild (Vannini & Vannini, 2016). Touristic mushing activities could emphasise connections to the entire ecology of wildness, as mushing and the presence of sled dogs contribute to the profiling of nature as an accessible and welcoming 'natural' space (Bertella, 2016). However, non-humans are usually excluded from being considered as subjects contributing to the touristic experience (Bertella, 2014).

Tourism consists not only of the consumption and production of experiences happening in isolation from other elements of life, but is also 'enacted in combination with many other "non-touristic" practices' (Ren et al., 2019: 4), involving practices of non-human beings as components (Granås, 2018). Practices of non-humans, such as habitual behaviours of animals, often conflict with human categorisations of spaces, objects and ideas, in order to protect their own safety and spatial boundaries (Buller, 2014; Lulka, 2004). Mushing as a touristic activity exists to produce experiences, but as a practice it is inseparable from the embodied nature of movement in human–sled dog encounters. Like walking, mushing is an embodied practice; it is not only about journeying from place to place but also about an open-ended movement 'having neither a point of origin nor any final destination' (Ingold & Vergunst, 2008: 2). As an interspecies practice, mushing has to be learned by dogs, mushers and, at least to some extent, tourists. It is the human's responsibility to adapt to the dogs' embodied presence in order to understand the dogs, to respect and trust their abilities and, ultimately, to comprehend the dogs and communicate with them in a way they can understand (Kuhl, 2011).

In addition to dog sledding, mushing comprises of versatile practices related to taking care of the dogs. Developing and maintaining the housing systems require a huge amount of work. The dogs live in tethers or kennels. For a tourist, a tether is usually the worst option as it seems more restrictive of the dogs' freedom of movement. These are 'sites where the material reality of the housing system itself combines the practices of human and nonhuman agents, with science, and with politics' (Bjørkdahl & Druglitrø, 2016: 1). Dogs and tourists may not be fully aware of each other's modalities, nor of human policies allowing or restricting these modalities, which may lead to unpleasant encounters for one or both parties. Sled dog and human participants in touristic mushing activities have to be able to adapt to each other's bodily presence in order to develop shared practices; otherwise, positive encounters between them would be impossible. Therefore, there is the potential for both participants to make a difference in the encounter in versatile and unexpected ways (Koski & Bäcklund, 2017). Coping under these conditions requires that each party copies all the possible modalities of movement employed by the other and, as a result, both parties come to perceive the environment in similar ways (Ingold & Vergunst, 2008: 11).

Sled-Dog Agency in Tourism

Within the framework of tourism, we might think that animals are forcibly involved in tourism activities without the opportunity to consent to their involvement. Animals may be both unable to fully understand the regulations of engagement in such activities and vulnerable to human desires (Dashper, 2019). Even if they are aware of these complexities, they lack the equal freedom to choose; this is because animals are subject to human values, and human values are likely to be self-interested under commercial pressure (Dashper, 2014). This is particularly significant in a global industry such as tourism. Tourism as a social institution is not meaningful to animals as such, but it nevertheless fundamentally influences their lives. In turn, tourists are increasingly aware of the issues pertaining to animal-based tourism. Through client pressure, animals affect both the institutional structures and the people working in tourism (Notzke, 2019).

Agency is often understood as a manifestation of free will and moral behaviour possessed exclusively by human beings. This understanding becomes troublesome, however, as beings, including non-human animals, can adapt to the behavioural standards of their groups. These encounters can even be broadened to include an interspecies sense of morality (McFarland & Hediger, 2009: 6). Despite the shared and reciprocal nature of interspecies agency, however, animals' capacity to act and co-habit with humans is usually asymmetric and unequal (Haraway, 2003). Nevertheless, animal agency can have effects beyond the bounds of the spaces where they are immediately present, as human–animal relations operate beyond the boundaries of physical proximity and animals 'are able to have an effect on humans at-a-distance' (Philo & Wilbert, 2000: 2). The ability to act, effect change or make a difference is multidirectional and does not come from individuals alone, but 'is engaged in relations' (Urbanik, 2012: 43).

Dog agency is difficult to define, as the social space of dogs is both inside and outside of human society and our understanding of dogs' consciousness and self-fulfilment is limited (Koski & Bäcklund, 2017). Dogs are seen as both 'man's best friend' with the capability for rational thought, and as objects for human values. Sled dogs in particular occupy a liminal position, as they reside on the boundaries of the domestic and the wild (Onion, 2009). However, the shared interspecies agency of dogs and humans can blur the status of the sled dog as hovering between that of a pet, a working animal and even a (semi-)wild animal (Jones, 2003). Agency is relational and situational and varies depending on the conditions of time, space, materiality, embodiment and relations to other agents and their positions (Buller, 2012; Philo & Wilbert, 2000). Dogs cannot manifest agency by engaging in two-way verbal communication with humans about their housing conditions or their lives in general. However, mutual understanding can occur in an embodied way through dogs and humans engaging in non-verbal communication and collaboration (Danby & Hannam, 2016). Through movement, animals enact, develop and communicate their agency to others, just as humans do. This creates new co-assemblages of movement (Buller, 2012). Hence, to recognise the agency of sled dogs it is more important to pay attention to the emerging set of knowledges and practices shared through body movement, than it is to consider the mutual intentionality or sense of a broader ontological construction between the human and non-human actors (Buller, 2012).

Sled dogs adapt to the conditions of touristic mushing but their presence reveals a contradictory space; these contradictions include questions of self-fulfilment, forms of interaction, political possibilities and restrictions in structural and everyday power relations (Koski & Bäcklund, 2017; Philo & Wilbert, 2000). The interspecies agency between humans and sled dogs shapes touristic practices materially and socially with innovative outcomes, and accordingly produces material impacts, which disrupt the touristic representations of places (Granås, 2018). As a result, the sometimes-unexpected visibility of animals creates shared spaces, where new meanings are created by both humans and animals (Lulka, 2004). These new meanings can disrupt the representation of wilderness as a pristine place for wildlife tourism (Vannini & Vannini, 2016), which can have effects on our understanding of wildlife tourism. In the next section, we introduce a future scenario, which proposes how sled dogs as agents could contribute to experiencing wildness in the context of wildlife tourism.

Designing Future Wildlife Tourism Experiences with Sled Dogs

Our aim in creating a future scenario for wildlife tourism is to find alternative means of experiencing and developing the human–sled dog encounter. For our design approach, we have combined two different design methods: service design to provide a process structure and framework for our design activities, and speculative design to provide us with future-oriented design tools.

Service design

Service design is a holistic field that looks at service systems in terms of the core principles of human centeredness and co-creation (Sanders & Stappers, 2008). In examining the needs of all agents in a service system, service design aims to provide value through creating service solutions that may contain both digital and physical elements (Miettinen & Koivisto, 2009). In multichannel service systems, the service environment itself also plays a significant role in creating a service experience. Designers can implement solutions inspired by nature by highlighting animal practices and spatial perceptions as components of wildness at different service stages.

By definition, a service is produced and consumed simultaneously. This is one basis for the view of value co-creation in a service encounter (Jaakkola & Alexander, 2014; Shostack, 1985), whether as human–human interaction or as an encounter between humans and non-humans. As the human-centred approach is dominant in service design, the views of non-human agents are often neglected. Nevertheless, many service encounters also include non-human agents such as machines or living entities such as animals; these agents should be considered in the value co-creation process of tourism services (García-Rosell *et al.*, 2019).

Our design activities take inspiration from studies concerning the service encounters between non-human and human agents in which animals are considered as active subjects in creating touristic experiences (Bertella, 2014). As a basis for our design process, we use the double diamond design process model (Design Council UK, 2015; Stickdorn *et al.*, 2018), a commonly used framework in service design. The process model is divided into four phases. The discovery phase (1) aims to collect information and to observe the conditions, activities and agents in the service in order to find out the needs and challenges that exist. The definition phase (2) is based on the findings from the previous phase and aims to define the design challenge that is then taken into the (3) ideation phase. The delivery phase (4) includes the final design of the service solution and its implementation in the service practice.

Speculative design

Speculative design is the practice of using creative methods to develop conceptual alternative futures and to imagine possible solutions, in order to identify design challenges. The approach aims to provide better understanding of current issues and to generate dialogue for future solutions and desires. Speculative design does not make predictions or forecasts. It instead provides alternatives without stating claims of truth (Dunne & Raby, 2013), thereby giving space for creativity without the boundaries of the current world. Speculative design can be used as a tool to explore futures that help us reflect on, understand and possibly alter current situations (Barbrook, 2007). Additionally, it incorporates elements of critical design such as questioning the underlying assumptions of life, avoiding taking ideas for granted and exploring alternatives (Dunne & Raby, 2013). As a design tool, speculative design is inspired by science fiction and allows designers the freedom of being detached from the world's current boundaries (Yeoman & Postma, 2014).

Design process

Based on the service design double diamond process model, a series of design activities were conducted by the authors. In the discovery phase, several video recordings of touristic mushing activities were used as empirical material for observation. The videos recorded a touristic summertime mushing experience that occurred in Finnish Lapland in 2019. The authors focused on observing the following: sled dog behaviour, interactions between sled dogs and mushers, the environment and the interactions between sled dogs and tourists. As a result, four main themes for ideation were defined: (1) the agency of sled dogs, (2) bodily movement, (3) the use of technology and (4) trends in wildlife tourism.

Using speculative design as a tool, 'what-if' speculation was geared towards ascertaining scenario drivers, needs and possible emotional aspects concerning a future scenario in which the sled dog is the centre of human–animal interaction in a wildlife tourism context. By applying this animal-centred approach, the following design drivers were chosen to guide the creation of the final future scenario:

- What if, in the future, the dog's needs and free will came before the customer's and service provider's perspective in mushing?
- What if tourism operators based their activities significantly on scientific research on sled-dog agency (e.g. movement in sled dog–human encounters)?
- What if dogs had the power to decide whether they want to interact with humans, or even to decide not to work at all?
- What if technological devices, such as cameras and phones, were not permitted at all on the sites where dog sledding activities take place?
- What if mushers and tourists were able to comprehend the emotions, vital statistics and preferences of dogs through human- and animal-embedded technology?
- What if only a limited group of people were able to participate in sled-dog activities?
- What if it was mandatory that tourism agencies educate tourists before permitting them to participate in sled-dog activities?

The design drivers were used as inspiration for ideating possible solutions that could exist in the kind of future that the drivers frame. Contradicting the usual human-centred approach, we tried to remain animal centred throughout the exercise in order to be free from our human assumptions as much as possible. Thus, we were able to create an alternative and speculative scenario for wildlife tourism. The resulting scenario is introduced in the following section.

Future Scenario: A Day in the Life of Alanis

In addition to the defined design drivers, we assumed a future in which the structure of the touristic mushing activities would not have changed drastically. Therefore, we were able to use the structure of an existing full-day sled-dog safari as the basis for our scenario, which represents a 24-hour period from the perspective of a sled dog. Through the utilisation of technology and the recognition of the importance of movement in the environment, the scenario was built to communicate a possible multisensory wildlife tourism experience (Wohlwill, 1966).

The character presented in the scenario is called Alanis, an alias for one of the dogs observed in the first design phase. Alanis is a female Alaskan husky who lives in Finnish Lapland, in a kennel together with another dog. At birth, a chip was inserted under her skin, allowing her vitals to be monitored and stored, including information on her moods, needs and preferences. Without making any claims about how the technology might work, we assume that it plays an important role in translating the needs of the dog towards humans, thereby amplifying her exercise of agency.

During her day, Alanis interacts with other dogs in her surroundings. In our scenario, we recognise that dog-to-dog communication is an important part of Alanis's life. Because this is a natural part of her behaviour as a dog, we do not want to intervene with this interaction by introducing any technological solution to this issue.

Other agents around Alanis are the mushers and the tourists who are participating in the mushing activity. During our observation, we realised that the tourists are often unable to form a personal connection with the dogs. The use of technologies such as phones is acting as a barrier between the humans and animals, preventing the tourists from being fully immersed in the moment. Therefore, in our scenario all technological devices from the tourists are banned. Instead, they are provided on their arrival with a small earpiece, which allows them to access the transmission of one dog's data and real-time information during the mushing activity. Through this technology, the tourist is able to better understand the behaviour and mood of Alanis, thereby forging a more personal connection with her.

Currently, mushers probably have the greatest knowledge of dogs' behaviours and needs. In our scenario, an earpiece and a technological contact lens provide real-time information pertaining to the dog's vitals, emotions and needs. The earpiece functions as a transmitter of the dog's data through voice. The lens adds augmented reality to the vision of the musher, showing alerts based on the vitals of an individual dog. Such technology pieces can also be used for receiving real-time weather information and navigation road maps, as well as alerts about the environment. Mushers are then able to use this information in their interactions with and guidance of the dogs. These fictional technological tools have been utilised in this study to create a scenario concerning a typical day in the life of Alanis during the tourist season. Figure 10.1 places Alanis as the key actor in a scenario for future wildlife tourism, illustrating clockwise the various activities she is involved in, as well as her interactions with other dogs and human participants (i.e. mushers and tourists).

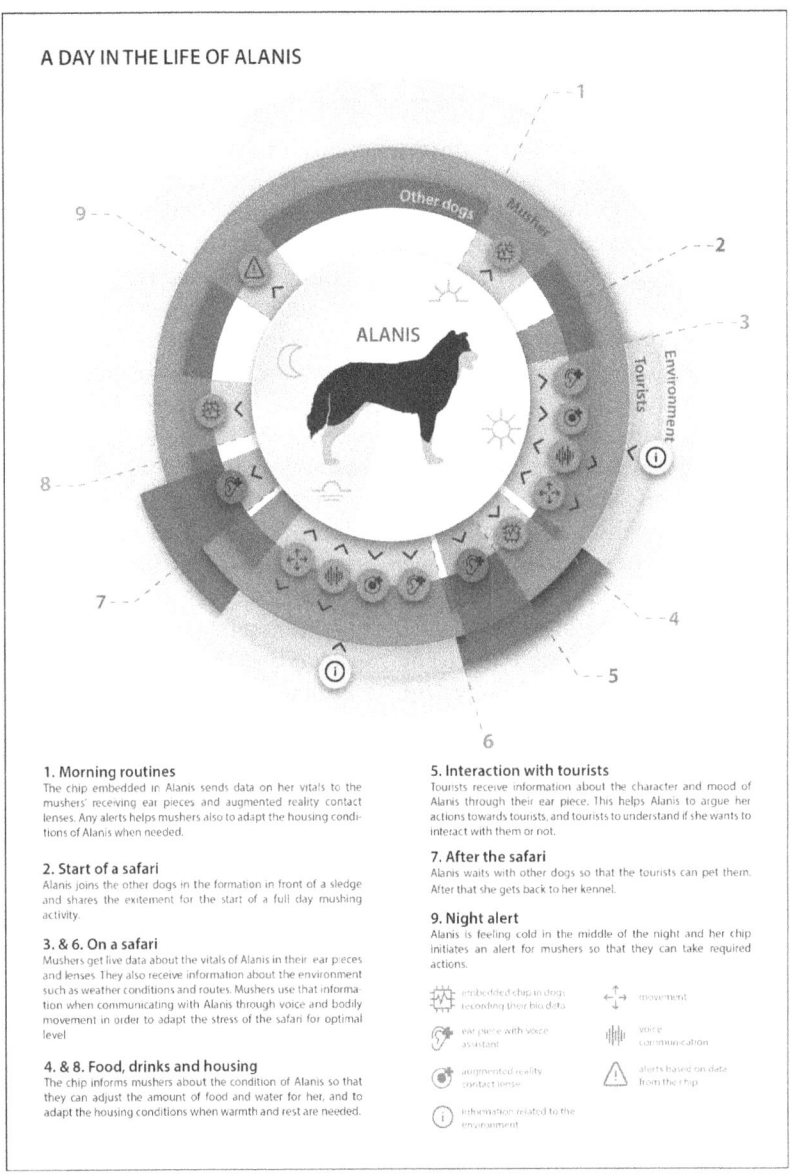

Figure 10.1 'A day in the life of Alanis': Future wildlife tourism experience from a sled dog's point of view

Discussion

Through the speculative scenario, we explored how sled dogs could be better acknowledged as agents in the interaction of components of wildness through the use of technological solutions that better enable the interspecies encounter to be experienced and interpreted.

The information generated by the future scenario is primarily based on the assumption that the role of the animal can be put in the centre of the value co-creation (García-Rosell *et al.*, 2019). By designing services in a way that moves beyond the limiting abstract symbols and icons of human language, and the overestimation of our mastery and agency, we can leave room for acknowledging the agency of other beings. In our scenario, we have demonstrated how utilisation of human- and animal-embedded technology can offer a better chance for animals to manifest their agency in tourism. Through technology, humans are better able to understand the needs and emotions of animals. Therefore, a service design supported by future technology could enable the interpretation of human–animal interactions and eventually make them more meaningful.

We have discussed that in the future humans may have more accurate and versatile means of engaging in mutual understanding with sled dogs, which will help with the acknowledgement of animal agency in tourism. We have also imagined that this possibility will impact how tourism experiences will be designed and lived. Recognising animals as agents and creating more possibilities to engage in mutual meaningful interaction also entails ethical implications, as the agency of animals impacts their own fate which oftentimes depends on human actions and desires (Notzke, 2019). If animals are to be considered as co-creators of experiences, we have to move away from a human-centred approach which neglects the capability of animals for agency. Future technologies might be a solution but they also offer the possibility of further complexity, as technological devices have agency and may themselves shape the human–animal encounter (Brown & Banks, 2015). The continuous presence of technology and engagement with complex information might therefore disrupt the unique embodied encounter between non-human animals and humans (Webber *et al.*, 2017). Still, technological advances create a compelling possible future for re-scaled, embodied interspecies encounters (Hodgetts & Lorimer, 2015: 289). As a result, we may have to radically change the ways in which we practice animal-based tourism – or even suspend them altogether.

Conclusions

In this chapter, we have explored how touristic mushing might contribute to envisioning wildlife tourism futures when sled-dog agency is acknowledged. In our speculative scenario, we placed a particular dog in the centre of a fictional future service and explored the moving encounters this dog generated with other agents during a typical day. These speculative encounters were made more meaningful through the use of future technology, which enabled non-human and human agents to better interpret and understand each other. In the context of touristic mushing, sled dogs are important agents since they contribute to the quality of the interaction among the components of wildness including humans and the environment. Indeed, components of wildness might include not only encounters with wildlife but also the embodied experience of our own animality (Curtin, 2010). The embodied experience is enabled by the close encounter with sled dogs and mediated by technological devices. Meaningful encounters can also lead to an educational experience if a tourist is able to sense the needs and emotions of the dog. As a result, shared spaces and new meanings can be co-created through human–animal interactions; this is of particular significance to the ethical dimension of animal-based tourism.

Through discussion of the concept of agency, we have drawn attention to encounters between humans and sled dogs in tourism and, more importantly, to the capacities of animals to be co-creators of tourism experiences. Non-human animals shape such encounters through their actions by bringing their own life histories and experiences to the encounters. We recognise that our future scenario does not erase the issue of human interpretation in terms of animal's actions, but we hope that it is a step forward in creating possibilities for more meaningful ways for animals and humans to interact in the context of tourism. In our view, the design of services should be concerned with finding ways to extend the mutual capacities of humans and non-human animals. More collaborative human–animal encounters might also contribute to a mutual development of tourism knowledge (Picken, 2018). Through concern for sled-dog agency in relation to wildlife tourism, we call for a closer investigation of differing spaces and practices of particular animalities in particular spaces of tourism. Notwithstanding that agency is a highly complex concept, it can bring new insights through the acknowledgement of animals as co-creators in the future planning of (wildlife) tourism.

References

Ballantyne, R., Packer, J. and Sutherland, L.A. (2011) Visitors' memories of wildlife tourism: Implications for the design of powerful interpretive experiences. *Tourism Management* 32 (4), 770–779.

Barbrook, R. (2007) *Imaginary Futures: From Thinking Machines to the Global Village.* London: Pluto Press.

Bentrupperbäumer, J. (2005) Human dimension of wildlife interactions. In D. Newsome, R.K. Dowling and S.A. Moore (eds) *Wildlife Tourism* (pp. 82–112). Clevedon: Channel View Publications.

Bertella, G. (2014) The co-creation of animal-based tourism experience. *Tourism Recreation Research* 39, 115–125.

Bertella, G. (2016) Experiencing nature in animal-based tourism. *Journal of Outdoor Recreation and Tourism* 14, 22–26.

Bjørkdahl, K. and Druglitrø, T. (2016) Animal housing/housing animals: Nodes of politics, practices and human–animal relations. In K. Bjørkdahl and T. Druglitrø (eds) *Animal Housing and Human–Animal Relations: Politics, Practices and Infrastructures* (pp. 1–14). Abingdon: Routledge.

Brown, K.M. and Banks, E. (2015) Close encounters: Using mobile video ethnography to understand human-animal relations. In. C. Bates (ed.) *Video Methods: Social Science Research in Motion* (pp. 95–120). Abingdon: Routledge.

Buller, H. (2012) 'One slash of light, then gone': Animals as movement. *Etudes Rurales* 189, 139–154.
Buller, H. (2014) Reconfiguring wild spaces: The porous boundaries of wild animal geographies. In G. Marvin and S. McHugh (eds) *Routledge Handbook of Human–Animal Studies* (pp. 39–53). Abingdon: Routledge.
Cookson, L.J. (2011) A definition for wildness. *Ecopsychology* 3 (3), 187–193.
Curtin, S. (2009) Wildlife tourism: The intangible, psychological benefits of human–wildlife encounters. *Current Issues in Tourism* 12, 451–474.
Curtin, S. (2010) What makes for memorable wildlife encounters? Revelations from 'serious' wildlife tourists. *Journal of Ecotourism* 9 (2), 149–168.
Danby, P. and Hannam, K. (2016) Entrainment: Human-equine leisure mobilities. In J. Rickly, K. Hannam and M. Mostafanezhad (eds) *Tourism and Leisure Mobilities* (pp. 27–38). Abingdon: Routledge.
Dashper, K. (2014) Tools of the trade or part of the family? Horses in competitive equestrian sport. *Society & Animals* 22 (4), 352–371.
Dashper, K. (2019) Moving beyond anthropocentrism in leisure research: Multispecies perspectives. *Annals of Leisure Research* 22 (2), 133–139.
Design Council UK (2015) The Double Diamond Design Process. See https://www.designcouncil.org.uk/news-opinion/design-process-what-double-diamond (accessed October 2019).
Dunne, A. and Raby, F. (2013) *Speculative Everything: Design, Fiction, and Social Dreaming*. Cambridge, MA: MIT Press.
Fennell, D.A. (2012) *Tourism and Animal Ethics*. Abingdon: Routledge.
Fiszdon, K. and Czarkowska, K. (2008) Social behaviours in Siberian huskies. Annals of Warsaw University of Life Sciences – SGGW. *Animal Science* 45, 19–28.
García-Rosell, J.C and Äijälä, M. (2018) Animal-based tourism in Lapland. In J. Ojuva (ed.) *Animal Welfare in Tourism Services: Examples and Practical Tips for the Well-Being of Animals Used for Tourism in Lapland* (pp. 10–24). Rovaniemi: Lapland University of Applied Sciences.
García-Rosell, J.C., Haanpää, M. and Janhunen, J. (2019) 'Dig where you stand': Values-based co-creation through improvisation. *Tourism Recreation Research* 44 (3), 348–358.
Granås, B. (2018) Destinizing Finnmark: Place making through dogsledding. *Annals of Tourism Research* 72, 48–57.
Haraway, D. (2003) *The Companion Species Manifesto*. Chicago, IL: Prickly Paradigm Press.
Higginbottom, K. (2004) Wildlife tourism: An introduction. In K. Higginbottom (ed.) *Wildlife Tourism: Impacts, Management and Planning* (pp. 1–15). Altona Vic: Common Ground Publishing.
Hodgetts, T. and Lorimer, J. (2015) Methodologies for animals' geographies: Cultures, communication and genomics. *Cultural Geographies* 22 (2), 285–295. doi: 10.1177/1474474014525114
Hughes, K. (2013) Measuring the impact of viewing wildlife: Do positive intentions equate to long-term changes in conservation behaviour? *Journal of Sustainable Tourism* 21, 42–59.
Ingold, T. and Vergunst, J.L. (2008) Introduction. In T. Ingold and J.L. Vergunst (eds) *Ways of Walking: Ethnography and Practice on Foot* (pp. 1–19). Abingdon: Routledge.
Jaakkola, E. and Alexander, M. (2014) The role of customer engagement behaviour in value co-creation: A service system perspective. *Journal of Service Research* 17 (3), 247–261.
Jones, O. (2003) 'The restraints of beasts': Rurality, animality, Actor Network Theory and dwelling. In P. Cloke (ed.) *Country Visions* (pp. 283–303). Harlow: Pearson Education.

Koski, L. and Bäcklund, P. (2017) Whose agency? Humans and dogs in training. In T. Räsänen and T. Syrjämaa (eds) *Shared Lives of Humans and Animals: Animal Agency in the Global North* (pp. 11–23). Abingdon: Routledge.

Kuhl, G. (2011) Human–sled dog relations: What can we learn from the stories and experiences of mushers? *Society & Animals* 19, 22–37.

Lorimer, J. (2007) Nonhuman charisma. *Environment and Planning D: Society and Space* 25 (5), 911–932.

Lulka, D. (2004) Stabilizing the herd: Fixing the identity of nonhumans. *Environment and Planning D: Society and Space* 22 (3), 439–463.

McFarland, S.E. and Hediger, R. (2009) Approaching the agency of other animals: An introduction. In S.E. McFarland and R. Hediger (eds) *Animals and Agency: An Interdisciplinary Exploration* (pp. 1–20). Leiden: Brill.

Miettinen, S. and Koivisto, M. (eds) (2009) *Designing Services with Innovative Methods*. Keuruu: Kuopio Academy of Design.

Notzke, C. (2019) Equestrian tourism: Animal agency observed. *Current Issues in Tourism* 22 (8), 948–966.

Onion, R. (2009) Sled dogs of the American North: On masculinity, whiteness and human freedom. In S.E. McFarland and R. Hediger (eds) *Animals and Agency: An Interdisciplinary Exploration* (pp. 129–155). Leiden: Brill.

Philo, C. and Wilbert, C. (2000) Animal spaces, beastly places: An introduction. In C. Philo and C. Wilbert (eds) *Animal Spaces, Beastly Places: New Geographies of Human–Animal Relations* (pp. 1–35). London: Routledge.

Picken, F. (2018) Knowing the aquatic other: Unleashing blackfish. In C. Ren, G.T. Jóhannesson and R. van der Duim (eds) *Co-creating Tourism Research: Towards Collaborative Ways of Knowing* (pp. 147–161). Abingdon: Routledge.

Ren, C., James, L. and Halkier, H. (2019) Practices in and of tourism. In L. James, C. Ren and H. Halkier (eds) *Theories of Practice in Tourism* (pp. 1–9). Abingdon: Routledge.

Sanders, E.B.N. and Stappers, P.J. (2008) Co-creation and the new landscapes of design. *Co-design* 4 (1), 5–18.

Shostack, G.L. (1985) Planning the service encounter. In J.A. Czepiel, M.R. Solomon and C.F. Suprenant (eds) *The Service Encounter: Managing Employee/Customer Interaction in Service Businesses* (pp. 243–254). Lexington, KY: Lexington Books.

Skibbins, J.C. (2015) Ambassadors or attractions: Disentangling the role of flagship species in wildlife tourism. In K. Markwell (ed.) *Animals and Tourism: Understanding Diverse Relationships* (pp. 256–273). Bristol: Channel View Publications.

Stickdorn, M., Hormess, M., Lawrence, A. and Schneider, J. (2018) *This is Service Design Doing: Using Research and Customer Journey Maps to Create Successful Services*. Sebastopol: O'Reilly Media Inc.

Urbanik, J. (2012) *Placing Animals: An Introduction to the Geography of Human–Animal Relations*. Lanham, MD: Rowman & Littlefield Publishers.

Vannini, P. and Vannini, A. (2016) *Wilderness*. Abingdon: Routledge.

Webber, S., Carter, M., Smith, W. and Vetere, F. (2017) Interactive technology and human–animal encounters at the zoo. *International Journal of Human-Computer Studies* 98, 150–168.

Wohlwill, J.F. (1966) The physical environment: A problem for a psychology of stimulation. *Journal of Social Issues* 22 (4), 29–38.

Yeoman, I. and Postma, A. (2014) Developing an ontological framework for tourism futures. *Tourism Recreation Research* 39 (3), 299–304.

11 The Future of Captive Animals and Tourism: The Zoo and Aquatic Cloning Centre 2070

Daniel William Mackenzie Wright

> The greatness of a nation can be judged by the way its animals are treated.
> Mahatma Gandhi

When we lose an animal species to extinction, a part of us is lost, faded into history. But what if we had the ability to give life back to the animals that over the years have become extinct? Or, with foresight, humans stored the DNA of animals on the brink, with the intention of prolonging their existence when their time eventually comes to a desperate end, via DNA cloning. The practice of cloning animals is arguably still a challenging issue, and interests across the globe can vary from taboo, ethically questionable to a welcomed scientific endeavour. In a previous study, the author introduced the idea of cloning animals for tourism (Wright, 2018). In the article, the author presented three scenarios in which tourism and cloning animals could exist in the future – Food Tourism 2070: A Japanese Restaurant Perspective; Sport Hunting and Tourism 2070: Designated Hunting Reserves in South Africa; and Safari Zoo Tourism 2070: Into the Wild, Cloning for Education and Conservation in the USA. It is the latter of these three scenarios, the idea of future zoos where cloning takes place for education and conservation, that is of interest to this chapter. In this chapter, the author aims to explore the possibility of adopting cloning technology in tourism in further detail. Cloning is a practice that already exists in society. Science continues to reach new levels within the field, as significant advances have been made since Dolly the Sheep in 1996. In the scenario from the author's previous study, he briefly considered the importance of education and preservation in future zoos. This chapter aims to explore the idea further, considering the evolution of zoos and aquariums in a society depleted of natural wildlife and reflecting on relevant consumer attitudes towards animals in captivity and animal cloning.

The importance of this study can be placed within Fahey and Randal's (1998) plausibility agenda. Here, the focus is on, 'plausible evidence should indicate that the projected narrative could take place (it is possible), demonstrate how it could take place (it is credible), and illustrate its implications for organisations (it is relevant)' (Fahey & Randal, 1998: 9). Applying a multidisciplinary and pragmatic approach, this study embraced views from qualitative and quantitative research, from positivistic to constructivists views (Lee, 2012; Tashakkori & Creswell, 2007). Importantly, the future is open to design ideas, and by exploring past, current and future knowledge, this chapter offers a narrative scenario that is fundamentally rooted in evidence. To do this, the author presents the future narrative scenario in a promotional material format. The promotional material is revealing the following: The Zoo and Aquatic Cloning Centre 2070. The promotional material offers a narrative of what a future zoo and aquarium could entail. Importantly, future zoos and aquariums offering touristic experiences where animal cloning is present will depend on the demand and desire of tourists to visit. Thus, this chapter raises difficult questions regarding attitudes and developments in animal cloning and tourism. The chapter initially presents a brief history of zoos and aquariums, and reflects on the question whether zoos and aquariums will still exist in the future. Subsequently, it presents the increasing loss of the natural environment as we are witnessing. Such reflections/considerations are the point of departure for the development of a future scenario about animal cloning in tourism. Such a scenario is presented and discussed in relation to relevant consumer attitudes. The concluding thoughts aim to provoke the readers' reflections: What do you think about captive wildlife? What about the possibility of cloning animals for tourism: would you visit such a tourism attraction?

A Brief History and Evolutionary Purpose of Zoos and Aquariums

> One would expect that an ancient institution such as the zoo would have long exhausted its selling powers, but the zoo continues to attract the masses.
> Braverman, 2011

The long history of zoos and aquariums has been well documented across the literature landscape. Zoos and aquariums have transitioned over time. Their evolution has focused on their purpose, their physical structure and layout (with consideration to both animal welfare and guest engagement) and, more widely, their relevance, role and responsibility to society and supporting our natural environment (Braverman, 2011; Mullan & Marvin, 1987, 1999; National Geographic Society, 1996–2019; Sylph, 2018). Looking back to their origins, different approaches were taken. From wealthy elite collecting animals from around the world for curiosity, leisure and pleasure, to more considered approaches where zoos and aquariums focused on animal and wildlife education and science (Kisling, 2000; Rothfels, 2002). The current oldest zoo was built in Tiergarten Schönbrunn in Vienna, Austria, opening its doors to the public in 1765 (Braverman, 2011). While the first zoos focused more on the separateness ideology, where human and nature was kept apart, in 1907 German Carl Hagenbeck opened the first bar-less zoo. Here, the focus was on the vision of presenting animals living more in nature, in unity with biological diversity (Mullan & Marvin, 1987, 1999): 'nothing more than unseen ditches were to separate wild animals from the members of the public' (Baratay & Hardouin-Fugier, 2004: 263). American landscaper Frederick Law Olmsted played a pivotal role in developing zoos that offered winding paths, wide vistas and picturesque locations with little artificial human interference. Olmsted believed that nature could offer psychic recreation to tired city workers (Braverman, 2011). The zoos would provide spaces where visitors could enjoy the benefits of the tranquil English garden concept. The notion was to be more with nature and moving beyond the idea of entertaining visitors to a more educational experience, offering natural history studies and cultural knowledge (Hanson, 2002). The physical environment continues to evolve as differences in practices continue, from caged approaches to more safari-like centres available to consumers, where education, entertainment, science and conservation are key characteristics. Today, greater emphasis is being placed on conservation, with zoos and aquariums increasingly identifying themselves as centres of education and conservation (Miller *et al.*, 2004). Patrick *et al.* (2007) note that many of these themes can be seen in their mission and visions, as current centres aim to be leaders in education and conservation.

Some authors do not believe that zoos and aquariums can offer opportunities for conservation, a key purpose of their existence. Safina (2018: 6–7) notes that, 'Zoos and aquariums are not necessarily miserable places for many kinds of animals to live. And on the other hand, being wild is no picnic. Life in a zoo can be longer, safer, and more comfortable. But zoos are basically dead ends. Captive breeding can occasionally help conservation. But captivity can never be conservation'. However, zoo conservation would very much contest this point. According to the Zoological Society of London (2019), good zoos do much more than just put animals on display for visitors. By breeding species at the risk of extinction in the wild, they play a significant role in conservation. Likewise, breeding aquatic animals for conservation is an important aspect of aquariums. For example, the Bristol Aquarium (2019), which is involved in breeding animals, states that 'captive breeding not only helps ease the pressure on wild populations but also allows more research to be done both on their behaviour and on ways to help safeguard their long-term future in the wild'.

While data is somewhat limited on global and regional visitor numbers and financial investment, the following stats do provide some insight into the industry:

- Annually more than 700 million people visited zoos and aquariums worldwide, as identified in a survey by Gusset and Dick (2011) conducted for the World Association of Zoos and Aquariums (WAZA), in collaboration with national and regional zoo and aquarium associations.
- US$350 million is spent annually on wildlife conservation, as reported by the world zoo and aquarium community.

The Association for Zoos and Aquarium (AZA) has 233 accredited zoos and aquariums in nine countries, and offers some useful statistics. According to AZA (1997–2019), its 2018 figures highlight that across all institutions, they have 800,000 animals in care, in accredited zoos and aquariums, 6,000 species and 1,000 threatened or endangered species.

- AZA spends $200 million yearly on conservation support projects and 115 reintroduction programmes (40+ for threatened or endangered species), supporting the reintroduction of species back into their natural habitats.

The following points relate to AZA accredited members in the United States:

- AZA accredited zoos and aquariums contributed more than $22.5 billion to the US economy in 2016.
- Supported up to 208,000 jobs.
- Welcomed more than 195 million visitors annually.
- From an informal science and education perspective, annually welcomed 50 million children visitors with families.
- 12 million student learners on field trips.
- 400,000 teachers trained in informal science education methods over the last decade. (AZA, 1997–2019)

The End of Zoos and Aquariums?

> Yet freedom is exactly what zoos deny their animal residents.
> Pierce and Bekoff, 2018: 43

Based on the aforementioned considerations, the following questions can be asked: How will zoos and aquariums evolve? Does future society have a place for zoos and aquariums? The answers to these questions might depend on how the sector and the public understand and engage in animal welfare and well-being. According to Kagan *et al.* (2018), 'The Detroit Zoological Society (DZS) has long challenged zoos' and aquariums' relative reluctance to acknowledge gaps in the welfare of exotic nonhuman animals in captivity and has facilitated efforts to recalibrate zoo and aquarium practices and policies accordingly'. Pierce and Bekoff (2018) recognised that zoo leaders purposefully attended animal welfare conferences, such as those in Detroit, to discuss current ethical issues. However, the issue is that, ethically, captivity means that animal freedom is not possible. The literature explores the ethical challenge on limiting the negative impacts on animal welfare because of captivity. Nevertheless, discussions often turn to the ethical issues arising in zoos and not on the actual ethics of zoos themselves (Pierce & Bekoff, 2018). Pierce and Bekoff (2018: 44) note that 'even the best welfare in the world will not make zoos ethically benign or even acceptable. What we need is a paradigm change. Zoos need to radically reform and likely will not look anything like today's zoos'. To dramatically change the current problems in zoos and the negative impacts on animal welfare, the authors recommend several reforms, including shutting down bad zoos as soon as possible, stopping captive animal breeding and using the science of animal cognition and emotion on behalf of animals. The reforms include (but are not limited to) practices such as killing healthy animals, captive breeding programmes, relocating animals and using science to further our understanding of animal cognition and emotion (Pierce & Bekoff, 2018: 45–47).

Focusing on the ethical dimension of animal captivity, Pierce and Bekoff (2018: 44) suggest that: 'one option is to ignore ethics and simply admit, unabashedly, that ethics is not that important. We can be upfront about the fact that we are exploiting animals, that we are causing them to suffer physically and mentally, that we do not know how to fix the innumerable problems at hand, or that we simply do not care enough about animal well-being to stop supporting and participating in zoos'. Such an option is in conflict with what is highlighted by The People for the Ethical Treatment of Animals (PETA, 2019) that supports people to become more informed about the behaviour and requirements of wild animals and the negative impacts that come about due to captivity. On the other hand, in a recent collaboration, World Animal Protection (2019) and Change for Animals Foundation (2019) surveyed more than 1200 zoos and aquariums that belong to the WAZA and claim to be the world's leading zoos and aquariums. The organisations stated that they visited 12 of the 'top' zoos and aquariums and from their visits they found that 'cruel and demeaning performances and activities, known to cause great physical and mental distress, are taking place' (Change for Animals Foundation, 2019). Their report states that they found 43% offered visitors the opportunity to pet the animals, 33% allowed tourists to walk or swim through the animals' enclosures, 30% used their animals in live shows, 23% allowed visitors to hand-feed the animals, while 5% allowed tourists to ride the animals, all of which fail to meet the standards set by WAZA, which states that modern zoos and aquariums should not participate in animal shows or interactive experiences where animals perform unnatural behaviours. The organisations call out to tourists to offer their support to 'Never visit the zoos and aquariums we have featured in this report, or any that offer cruel animal attractions, and ask your friends and family not to either'; and 'Email WAZA directly demanding they stand up for the animals suffering in WAZA member zoos' (World Animal Protection, 2019: 18). The aforementioned organisations are asking tourists to boycott the zoos and aquariums and to ensure standards are met. Harry Eckman, director for Change for Animals Foundation, said, 'UK tourists can make a stand by not visiting or supporting these venues' (Manger, 2019). While the ethical and standards debates continue, what cannot be discounted is the current popularity of zoos and aquariums in society (as highlighted above with current visitor numbers). It would suggest that they will continue into the future and, if anything, their purpose and physical appearance will change as consumer expectations and demands continue to shine a light on the negative impacts on animals kept in captivity. Currently, there is little evidence of tourists' boycotting zoos and aquariums in any significant numbers.

Tourist motivations for visiting zoos and aquariums is a somewhat complex question to answer, as reasons are wide, varied and often individually subjective. Novelty seeking reasons could include the opportunity to see exotic animals on home soil. Zoos are essentially living museums, offering the opportunity to escape our everyday lives. Parents take children to educate them, to develop their understanding of animals and habitats or for entertainment purposes such as offering children the opportunity to bring their picture book and movie characters to life, to touch the animals and, in some cases, to ride or swim with them. By visiting zoos, humans can appreciate 'the distinctive quality of combining many kinds of animals along with a natural environment' (Sakagami & Ohta, 2010: 129). The human relationship with animals and nature is an area that continues to be studied. The biophilia hypothesis (also called BET) was popularised by Wilson (1984); it is the idea that humans have an innate tendency to seek out connections with nature and other forms of life. Our empathy and attitudes towards animals and the environment vary greatly across societies and cultures. Research suggests that our social upbringing will impact on our predisposed desires to seek out, or not, the company of animals (Jacobson *et al.*, 2012). This arguably raises some difficult issues; importantly, will people boycott zoos and aquariums for poor practice if they have little interest in the livelihoods of the animals?

Becoming Extinct: Animals and Their Habitats

> Civilisations are going to collapse and much of nature will be wiped out to extinction if humanity doesn't take urgent action on climate change, Sir David Attenborough has warned.
> Griffin, 2018

The scientific and academic community frequently informs us of the growing number of species that are at risk of extinction and the gradual depletion and changes in the natural environment on which species depend. Significantly, reports often highlight human behaviour as a key driver that continues to impact on the decline of wildlife populations globally. Importantly, however, these are declines, not yet extinctions according to Professor Ken Norris (Director of Science at Zoological Society of London) who stresses that this should be a wake-up call to encourage efforts to promote the recovery of animal populations (WWF, 2016). A call that would require a significant change in our attitudes towards the environment and our consumption of it. There is no doubt that animal extinction continues, as does the destruction of our (their) natural habitats. A recent report by the World Wildlife Fund (WWF, 2018), the world's leading independent conservation organisation, suggested that 'plummeting numbers of mammals, reptiles, amphibians, birds and fish around the world are an urgent sign that nature needs life support. Our Living Planet Report 2018 shows population sizes of wildlife decreased by 60% globally between 1970 and 2014'. The WWF (2018) states that, our 'current efforts to protect nature are not ambitious enough to match the scale of the threat our planet is facing. We are calling for a new global deal for nature and people to halt wildlife decline and tackle deforestation, climate change and plastic pollution'. The Agence France Presse (AFP, 2019), who obtained a leaked draft report by the UN, suggested that up to 1 million species face extinction due to human influence, cataloguing how 'humanity has undermined the natural resources upon which its very survival depends'. If the decline of our species is not through natural habitat destruction, then it is through voluntarily eating them out of extinction, as highlighted by Sawa (2019), in an article entitled Deadly Appetite: 10 Animals We are Eating into Extinction. Likewise, Cimons (2019) suggests that humans have a long history of killing large animals, and that we continue to eat them into extinction. And the result of all this, the extreme outcome due to 'a "biological annihilation" of wildlife in recent decades' is a sixth mass extinction facing humanity (Ceballos et al., 2015). According to Pyron (2017), associate professor of biology at the George Washington University, earth witnesses mass extinctions every 50–100 million years, and in the process, such events wipe out up to 95% of all species. Pyron (2017) further notes that climate scientists have warned us of our impacts on altering the planet, leaving us with little food, water and devastating weather patterns.

A recent report (compiled by 145 expert authors from 50 countries over the past three years, with inputs from another 310 contributing authors) has been presented by the Intergovernmental Science-Policy Platform on Biodiversity and Ecosystem Services (IPBES, 2019). This report offers a global assessment of nature by drawing on 15,000 reference materials. The findings focus on changes over the past five decades and highlight how impacts on our natural environment are related to economic development. In short, the report states that one million plant species and animals are threatened with extinction. Declining numbers of natural habitats and animals are occurring at alarming speeds never previously witnessed. This is due to our consumption of resources, particularly energy and food. Humans are the dominant species of the planet, so should we hold a level of responsibility? Certainly, future generations could look back and question our actions. Interestingly, through technological advancements, today's society will have provided future societies with a powerful capability, the ability to clone.

Future Scenario: The Zoo and Aquatic Cloning Centre 2070

> Organisms which are produced by asexual reproduction are genetically identical to one another, and in nature, many organisms, produce clones through asexual reproduction. Cloning is the process of producing genetically identical individuals of an organism, it can be done naturally, or humans can do it artificially.
>
> National Human Genome Research Institute, 2020

To date, cloning is not seen as a viable conservation strategy, but some researchers continue to remain optimistic that it has the potential to help threatened species in the future (Jabr, 2013). Pasqualino Loi from the University of Teramo in Italy was part of a team that successfully cloned an endangered mouflon sheep in the early 2000s, and he observes that 'once cloning of endangered animals is properly established, it will be a very powerful tool' (see Jabr, 2013). Humans clone animals, and with further improvements in cloning technology in the coming decades, cloning will likely continue and become more advanced (see Wright [2018] for an overview of animal cloning and ethical debates). In 2017, San Diego Zoo Institute for Conservation Research (SDZICR, 2017) established 'frozen zoos'. Two of the eight SDZICR strategic areas highlighting the purpose of frozen zoos are 'conservation genetics: sustaining and restoring genetic diversity through bioresource banking and research' and 'reproductive sciences: applying innovative science and technology to enhance reproduction'. These frozen zoos will allow future generations access to the genetic material of extinct species.

As noted above, the current trend for zoos and aquatic centres is towards species conservation through breeding programmes, so will cloning be an advanced form of current breeding programmes? Figure 11.1 presents an insight into the potential future zoo and aquarium, showing

THE WILDLIFE STORY: A GUIDED EXPERIENCE FOR VISITORS

Visitors at the centre enjoy a guided experience, a journey through the history of animal wildlife. Using the latest developments in interactive technology, visitors can tailor their experiences throughout the visit. Our mission is to offer a day that educates and entertains all our guests.

THE HISTORY CENTRE

When visiting the history centre, visitors will be educated on the origins and evolutionary cycle of animal wildlife. The past is brought to life through high-tech innovations, including augmented reality biospheres, virtual reality worlds and robotic animals. Visitors will be able to explore how ecosystems, natural environments and animal habitats changed and even perished over time, and the impact this had on the gradual depletion and extinction of species.

SCIENCE AND EDUCATION CENTRE

Continuing the journey, the science and education centre focuses on humanities capabilities in cloning technology. Here visitors will be educated on the origins of animal cloning science, exploring the developments in biotechnology and the process of producing genetically identical organism through natural and artificial methods. The centre also raises awareness to many of the ethical and moral debates that transpired throughout the course of time, as cloning gradually become socially accepted and widely practiced.

OBSERVATION CENTRE

The next stage in the journey offers visitors a lifetime experience, the chance to observe cloned animals in their natural environments. The centre offers visitors the chance to see the specialist conservation work being carried out, how science is giving life back to previously extinct species. The center provides a cloning breeding programme to bring back to life extinct species and continue to support endangered species. The centre has recreated the natural environments that replicate the major biomes of the world, aquatic, desert, forest, grasslands, and the tundra. Here visitors get a complete immersive experience. Pre-book for the opportunity to swim with cloned animals in our aquatic habitat pools.

FUTURE CONSERVATION, PRESERVATION AND ADVANCEMENTS CENTRE

The final centre provides visitors with an opportunity to glimpse into the future. Visitors will be educated on the changes taking shape in the natural environment and potential issues for animal wildlife. Future predictions explore worst and best-case scenarios and humanities role in shaping these futures. Future technologies and developments in cloning and biotechnology expose visitors to the vision of the park, and it role in conservation and preservation and how breeding through cloning techniques are supporting the reintroduction of animal life into natural environments.

THE GIFT SHOP

The park journey ends at the gift shop. Here visitors can purchase a range of souvenirs, from animal toys, educational books and photographs of their day out across the visitor center. There is also range of cloned food delicacies that can be purchased. Here visitors can pick up some cloned meat and even pre-order the cloning of an animal, which will eventually be delivered to them in consumption form. The ultimate gift, is the opportunity to clone their own animal. Here visitors have an opportunity to sign up to the 'clone your own' initiative. Visitors can sign up to cloning their own pets. Through various methods (either by taking their pet to the center or via the do it at home kit) visitors can pay for their pet's genetic material to be stored in the centers frozen zoo, ready for potential cloning in the future.

Figure 11.1 Future zoo and aquarium promotional material

some possible promotional material offered by such a centre. The focus of the centre is to inform tourists about animal cloning, its origins, its purposes and its benefits, and to offer consumers a unique purchasing opportunity. Visitors to this centre are guided through an experience, a journey that takes them through the History Centre, the Science and Education Centre, the Observatory Centre, the Future Conservation, Preservation and Advancements Centre and, finally, the Gift Shop. The promotional material offers a brief narrative further explaining the experience and purpose of each section of the centre and the potential visitor experiences on offer.

Consumer Attitudes Towards Animal Cloning in the Future

Without the collection of primary data, it is somewhat difficult to suggest how current and future tourists will perceive a cloning centre such as that presented above. Even today, not all individuals accept and visit zoos and aquatic centres. Thus, if such an attraction existed tomorrow or in the year 2070, it is likely that opposing societal and cultural attitudes would continue. The above scenario purposefully presents a range of cloning experiences to tourists, from more traditional motives of education and science, to more contemporary purposes, such as conservation and preservation, while also offering the idea that tourists could invest in the opportunity to clone their own pets. The aim was to consider the full scope of animal cloning, from intangible and tangible consumer experiences. The scenario aims to entice readers to question such a scenario. Broader debates, such as the purpose and motives of animal cloning now and in the future, are raised. Readers should also consider their potential role as consumers; are you empathetic towards animals and their natural habitats? Would you visit such an attraction if it existed today? What would you expect from such an attraction? Are aspects of the scenario unsettling, such as cloning one's pet? Why? Challenging and contested yes, but practices such as pet cloning (Duncan, 2018; Sheridan, 2018) and cloning food for consumption (DeFazio, 2017; Tanne, 2008; The Gallup Organization, 2008) already exist. Maybe this dilemma is based on the readers' current attitudes towards the destruction of natural environments and the growing extinction of species. With increasing media focus, do we feel individually or collectively responsible? Humans have only been around for 200,000 years, a tiny blip in the 4.5 billion years of our planet's history. Yet, we have had a greater impact on the Earth than any other species. All over the world, we are cutting down forests, using too much river water, choking our oceans with plastic and pushing many animals to extinction (WWF, 2018). For some, such as Pyron (2017), 'yes, we have altered the environment and, in doing so, hurt other species. But we are a part of the biosphere just like every other creature, and our actions are just as volitional, their consequences just as natural. Conserving a species we have helped to kill off, but on which we are not directly dependent, serves to discharge our own guilt, but little else'. Seriously, little else? Not even human curiosity, scientific exploration, shared responsibility?

Concluding Thoughts

This chapter set out to explore and offer an original insight into a potential future zoo and aquarium and how animal cloning could become a key practice of the attraction. The idea is presented to the reader in a promotional material format. A future centre where visitors would be educated on cloning technology and science, animals and conservation and the chance to clone their own pet. In so doing, it aims to capture the full potential of cloning, from intangible visitor experiences to tangible consumption. Importantly, the scenario aims to raise awareness and debate around contested topics, such as animal cloning and the future of zoos and aquariums. The ideas presented here should be contested, and further research should continue to explore the relationship between animal cloning, consumer attitudes and the role of zoos and aquariums.

However, the message here is also one of individual personal reflection, looking into the future allows us to also consider the present. According to Kagan *et al.* (2018) captive wildlife institutions should adopt and embrace themselves as organisations that focus on individual animal well-being. This approach would ensure that such institutions are driven by education and conservation, with the intention of informing millions of visitors on the importance of respecting 'animals in captivity as individuals who experience the world in species-specific and individually unique ways' (Kagan *et al.*, 2018: 59). Not all share similar sentiments. Pyron (2017) suggests that 'the only reason we should conserve biodiversity is for ourselves, to create a stable future for human beings'. This ideology would likely fuel future zoos and aquatic centres to clone animals for our own desires, for the consumption of food or to bring a dead pet back to life; human gratification above everything else. Maybe the bad practices of zoos should be driven out by good zoos. As noted by Safina (2018), 'good zoos' through professional organisations should aim to target zoos conducting bad practice by forcing them to change their approach or to close. It remains unclear how educational initiatives affect visitor behaviour (Ogden & Heimlich, 2009) and what actual impacts financial expenditures have on influencing conservation efforts (Ferraro & Pattanayak, 2006). Nevertheless, the large participant numbers and the conservation money that is spent suggest that zoos and aquariums have an active role to play in educating society and supporting wildlife conservation (Gusset & Dick, 2011; Zimmermann *et al.*, 2007). In the future, just like the past, and similarly today, a great number of different centres, zoos, aquariums and safari-like experiences could be operating. The visitor economy has the potential to continue offering consumers a variety of different purpose-built locations in which captive animals exists. The limits to our imagination are often our own ethical frameworks, which vary greatly across societies. Exploring the future is necessary, and depictions like the one presented here provide the parameters in which not only to comprehend potential futures, but also to question our present behaviours.

In wildness is the preservation of the world. (Thoreau, 1862)

References

Agence France Presse (AFP) (2019) One million species risk extinction due to humans: Draft UN report. See https://www.afp.com/en/news/15/one-million-species-risk-extinction-due-humans-draft-un-report-doc-1fu6ad1 (accessed April 2019).

Association for Zoos and Aquarium (AZA) (1997–2019) Zoo and aquarium statistics. See https://www.aza.org/zoo-and-aquarium-statistics (accessed April 2019).

AZA (1997–2019) Strategic plan: We are AZA. See https://www.aza.org/strategic-plan (accessed February 2019).

Baratay, E. and Hardouin-Fugier, E. (2004) *Zoo: A History of Zoological Gardens in the West*. London: Reaktion.

Braverman, I. (2011) Looking at zoos. *Cultural Studies* 25 (6), 809–842.

Bristol Aquarium (2019) Captive breeding and conservation for our aquatic life. See https://www.bristolaquarium.co.uk/captive-breeding/ (accessed May 2019).

Ceballos, C., Ehrlich, P.R., Barnosky, A.D., Garcia, A. Pringle, R.M. and Palmer, T.M. (2015) Accelerated modern human-induced species losses: Entering the sixth mass extinction. *Science Advances* 1 (5), 1–5.

Change for Animals Foundation (2019) Help end wild animal abuse in 'top' zoos and aquariums all over the world by emailing of CEO of the World Association of Zoos and Aquariums (WAZA). See https://www.changeforanimals.org/help-end-animal-abuse-in-top-zoos (accessed August 2019).

Cimons, M. (2019) We are eating large animals into extinction. See https://www.popsci.com/eating-large-animals-into-extinction (accessed April 2019).

DeFazio, F. (2017) Are we eating cloned meat? See https://www.scientificamerican.com/article/are-we-eating-cloned-meat/ (accessed May 2019).

Duncan, D.E. (2018) Inside the very big, very controversial business of dog cloning. See https://www.vanityfair.com/style/2018/08/dog-cloning-animal-sooam-hwang (accessed May 2019).

Fahey, L.R. and Randal, R. (1998) What is scenario learning. In L. Fahey and R. Randall (eds) *Learning from the Future: Competitive Foresight Scenarios* (pp. 3–43). New York: Wiley.

Ferraro, P.J. and Pattanayak, S.K. (2006) Money for nothing? A call for empirical evaluation of biodiversity conservation investment *PLoS Biology* 4 (4), 482–488.

Griffin, A. (2018) David Attenborough says civilisation will collapse if humanity doesn't take action on global warming at climate change talks. See https://www.independent.co.uk/environment/david-attenborough-climate-talks-cop-24-poland-global-warming-civilisations-collapse-a8664856.html (accessed April 2019).

Gusset, M. and Dick, G. (2011) *The Global Reach of Zoos and Aquariums in Visitor Numbers and Conservation Expenditures*. Wiley Online Library.

Hanson, E. (2002) *Animal Attractions: Nature on Display at Zoos*. Princeton, NJ: Princeton University Press.

Intergovernmental Science-Policy Platform on Biodiversity and Ecosystem Services (IPBES) (2019) Media release: Nature's Dangerous Decline 'Unprecedented'; Species Extinction Rates 'Accelerating'. See https://www.ipbes.net/news/Media-Release-Global-Assessment (accessed May 2019).

Jabr, F. (2013) Will cloning ever save endangered animals? See https://www.scientificamerican.com/article/cloning-endangered-animals/ (accessed May 2019).

Jacobson, K.C., Hoffman, C.L., Vasilopoulos, T., Kremen, W.S., Panizzon, M.S., Grant, M.D., Lyons, M.J., Xian, H. and Franz, C.E. (2012) Genetic and environmental influences on individual differences in frequency of play with pets among middle-aged men: A behavioral genetic analysis. *Anthrozoös* 25 (4), 441–456.

Kagan, R., Allard, S. and Carter, S. (2018) What is the future for zoos and aquariums? *Journal of Applied Animal Welfare Science* 21 (1), 59–70.

Kisling, V.N. (2000) *Zoo and Aquarium History: Ancient Animal Collections to Zoological Gardens*, Boca Raton, FL: CRC Press.

Lee, M. (2012) *Knowing Our Future: The Startling Case for Futurology*. Oxford: Infinite Ideas Limited.

Manger, W. (2019) Calls for tourists to boycott zoos that exploit animals for demeaning entertainment. See https://www.mirror.co.uk/news/uk-news/calls-tourists-boycott-zoos-exploit-18796652 (accessed August 2019).

Miller, B., Conway, W., Reading, R.P., Wemmer, C., Wildt, D., Kleiman, D., Monfort, S., Rabinowitz, A., Armstrong, B. and Hutchins, M. (2004) Evaluating the conservation mission of zoos, aquariums, botanical gardens, and natural history museums. *Conservation Biology* 18 (1), 86–93.

Mullan, R. and Marvin, G. (1987, 1999) *Zoo Culture*. Baltimore, MD: University of Illinois Press.

National Geographic Society (1996–2019) Zoo. See https://www.nationalgeographic.org/encyclopedia/zoo/ (accessed February 2019).

National Human Genome Research Institute (2020) *Cloning Fact Sheet*. See https://www.genome.gov/about-genomics/fact-sheets/Cloning-Fact-Sheet (accessed September 2020).

Ogden, J. and Heimlich, J.E. (2009) Why focus on zoo and aquarium education? *Zoo Biology* 28, 357–360.

Patrick, P.G., Matthews, C.E., Ayers, D.F. and Tunnicliffe, S.D. (2007) Conservation and education: Prominent themes in zoo mission statements. *Journal of Environmental Education* 38 (3), 53–60.

PETA (2019) Zoos: An idea whose time has come and gone. See https://www.peta.org/issues/animals-in-entertainment/zoos/ (accessed August 2019).

Pierce, J. and Bekoff, M. (2018) A postzoo future: Why welfare fails animals in zoos. *Journal of Applied Animal Welfare Science* 21 (1), 43–48.

Pyron, A.R. (2016) We don't need to save endangered species. Extinction is part of evolution. See https://www.washingtonpost.com/outlook/we-dont-need-to-save-endangered-species-extinction-is-part-of-evolution/2017/11/21/57fc5658-cdb4-11e7-a1a3-0d1e45a6de3d_story.html (accessed August 2019).

Rothfels, N. (2002) *Savages and Beasts: The Birth of the Modern Zoo*. Baltimore, MD: John Hopkins University Press.

Safina, C. (2018) Where are zoos going – or are they gone? *Journal of Applied Animal Welfare Science* 21 (1), 4–11.

Sakagami, T. and Ohta, M. (2010) The effect of visiting zoos on human health and quality of life. *Animal Science Journal* 81 (1), 129–134.

San Diego Zoo Institute for Conservation Research (SDZICR) (2017) Mission and purpose. See http://institute.sandiegozoo.org/who-we-are/mission-purpose (accessed May 2019).

Sawa, D.B. (2019) Deadly appetite: 10 animals we are eating into extinction. See https://www.theguardian.com/food/2019/apr/03/deadly-appetite-10-animals-we-are-eating-into-extinction (accessed April 2019).

Sheridan, K. (2018) Pet cloning is not just for celebrities anymore. See https://phys.org/news/2018-03-pet-cloning-celebrities-anymore.html (accessed May 2019).

Sylph, A. (2018) Artefact of the month: The Fish House at ZSL London Zoo – the first public aquarium. See https://www.zsl.org/blogs/artefact-of-the-month/the-fish-house-at-zsl-london-zoo-the-first-public-aquarium (accessed February 2019).

Tanne, J.H. (2008) FDA approves use of cloned animals for food. See https://www.ncbi.nlm.nih.gov/pmc/articles/PMC2213849/ (accessed May 2019).

Tashakkori, A. and Creswell, J.W. (2007) Exploring the nature of research questions in mixed methods research. *Journal of Mixed Methods Research* 1 (3), 207–211.

The Gallup Organization (2008) Europeans' attitudes towards animal cloning, Analytical Report. See http://ec.europa.eu/commfrontoffice/publicopinion/flash/fl_238_en.pdf (accessed May 2019).

Thoreau, H.D. (1862) Walking. *The Atlantic Magazine*. See https://www.theatlantic.com/magazine/archive/1862/06/walking/304674/ (accessed May 2019).

Wilson, E.O. (1984) *Biophilia*. Cambridge, MA: Harvard University Press.

World Animal Protection (2019) The show can't go on: Help end wild animal abuse in 'top' zoos and aquariums. See https://www.worldanimalprotection.org/take-action/waza (accessed August 2019).

Wright, D.W.M. (2018) Cloning animals for tourism in the year 2070. *Futures* 95, 58–75.

WWF (2016) *Living Planet Report 2016*. See https://www.wwf.org.uk/updates/landmark-report-shows-global-wildlife-populations-course-decline-67-cent-2020 (accessed February 2019).

WWF (2018) A warning sign from our planet: Nature needs life support. See https://www.wwf.org.uk/updates/living-planet-report-2018 (accessed February 2019).

Zoological Society of London (2019) How breeding programmes work. See https://www.zsl.org/education/how-breeding-programmes-work (accessed May 2019).

Zimmermann, A., Hatchwell, M., Dickie, L.A. and West, C. (2007) *Zoos in the 21st Century: Catalysts for Conservation?* Cambridge: Cambridge University Press.

12 Will Cryptogovernance Save the Wildlife Tourism Commons?

David Lusseau

Wildlife has become a key asset in tourism to both attract visitors to a wide variety of destinations as well as help communities to derive added value from tours (Twining-Ward et al., 2018; World Tourism Organization, 2015). Wildlife can be used in a variety of ways for tourism and recreation. Over the past 50 years, traditional extractive activities, such as trophy hunting, have been overtaken by what has been classified traditionally as non-consumptive activities, including seeking interactions in a natural setting and wildlife watching (MacMillan & Phillip, 2008). Here, I will focus on the latter activities because of the management challenges they present.

Wildlife and, more generally, nature-based tourism suffer from the dilemma that in most instances we cannot assign a property right to the wildlife used; it is a common good (Pirotta & Lusseau, 2015). For decades, this was not perceived as a problem because there was no evidence that when an operator or a tourist used wildlife, it restricted the access or opportunities of others to use the same wildlife. We now know that this is not the case and indeed the International Union for Conservation of Nature (IUCN) Red List now classifies tourism and recreation as a conservation threat to 5930 species (Higham et al., 2016; Lusseau & Mancini, 2018). Tourism can displace wildlife, including animals and plants, through its infrastructural footprint and when a significant proportion of a species range is affected, this causes a conservation threat. Tourism interactions can also disturb the daily lives of individual wildlife which adapt to those disturbances by perturbing their physiological and/or behavioural functions. Repeated exposure to tourism interactions can therefore affect the health of individual wildlife and subsequently decrease their demographic contributions through survival and reproduction (Amo et al., 2006; Ellenberg et al., 2007; Lusseau et al., 2006). These disturbances can also deteriorate habitats of high conservation value (Reed & Merenlender, 2008).

These observed wildlife tourism impacts have therefore led to repeated calls to treat all wildlife tourism as consumptive (Higham et al., 2016; Meletis & Campbell, 2007). From a management perspective, it means that the repeated use of wildlife by some operators and tourists will restrict the ability of others to use the same individuals/species either because they have been displaced (Lusseau, 2005) or because the population size has been decreased by repeated exposure of the population to disturbances (Pirotta et al., 2018). We, therefore, face a potential tragedy of commons situation which is notoriously difficult to govern sustainably (Ostrom, 2009; Ostrom et al., 1999).

In this chapter, I will explore how blockchain technology could be used to solve the key issue in many wildlife tourism conundra which is rights allocation for each time wildlife is used. To discuss the future advantages of this approach, I will use a scenario based on a typical wildlife tourism venture such as whale watching where a wildlife asset is free for all to use but the cumulative tourism use we make of it reduces wildlife density. In this scenario, all operators can use the same wildlife at any time and do not know on any given day how many wildlife encounters take place. This represents the setting of a typical wildlife tourism destination (Higham et al., 2009). I will explore the benefits of tokenising wildlife interactions in such a scenario, as well as describe other benefits that can be derived from a tokenised wildlife which apply more broadly to all wildlife use. Finally, I will introduce cryptogovernance approaches that may be required to avoid potential abuses of tokenised wildlife which could lead again to overexploitation.

Property Rights in Wildlife Tourism

In a classical extractive activity, such as hunting, rights are easy to define because they are directly associated with individual animals (the kill) and we can directly manage the assignment of the number of individual animals that users can extract from a population. In the case of viewing activities, it is the amount of time we spend with individual wildlife that restricts the ability of others to 'use' the same wildlife. In addition, some individuals will be more sensitive than others to tourism disturbance. Finally, disturbances occurring during sensitive activities (e.g. foraging) are more likely to bear greater health and demographic consequences than if they occurred during less sensitive activities (e.g. when animals are travelling) (Lusseau, 2014). Therefore, not only is the estimation of the total amount of interactions an animal population can sustain complex, but also the assignment of rights is virtually intractable.

In some locations, this problem has been addressed by assigning private property rights to whole wildlife populations, by literally fencing in the 'assets' on private land (Baum et al., 2017; Cousins et al., 2010;

Kinnaird & O'Brien, 2012). However, in many instances this approach is not possible, particularly for species – such as many of the charismatic species that are key attractors in wildlife tourism – that roam over large cross-boundary regions. Management issues also emerge regardless of property right assignment when countries do not have strong governance and agents cannot always be trusted (Baynham-Herd *et al.*, 2018). We have tended to approach the problem of wildlife tourism management either as a non-problem (non-consumptive activity) or as a classical wildlife management problem. In the former, we manage welfare aspects of the interactions to ensure that tourism is responsible, that is, adhering to some socially agreed best practices. In the latter, we manage the maximum amount of interactions a population can sustain considering that all interactions, regardless of quality, have the scope to elicit some level of wildlife 'extraction'. While this estimation process is complicated, it is not impossible (Christiansen & Lusseau, 2015; Pirotta *et al.*, 2015) and therefore we can define a total amount of time animals should spend interacting with tourists and manage interactions by blocking time or space annually (Lusseau & Higham, 2004; Ngoprasert *et al.*, 2017; Tyne *et al.*, 2015). From the user perspective, the problem is more complicated as we try to define how space or time are allocated among users and how to manage this allocation process (Mancini *et al.*, 2017; Pirotta & Lusseau, 2015).

This opens a range of management challenges which largely emerge because we cannot process each usage right allocation transaction separately and in time to determine when sustainability limits are reached. That is, when tourism operations reach *a priori* defined limits of acceptable change for their impacts on the wildlife resource and the community hosting the industry (Higham *et al.*, 2009; Mancini *et al.*, 2017). A company will operate several trips every day and there are typically multiple companies operating at a destination. Some companies may not use wildlife on every tour and, in addition, ad hoc recreational activities, which cannot be 'licensed', can use wildlife at the destination as well. For example, for a mature wildlife destination such as the Moray Firth (Scotland), about 50 dolphin usage transactions would have to be processed daily for tour operators alone. A centralised management authority cannot maintain a licensing scheme that can cope with this dynamic environment without trusted full collaboration of users and/or heavy policing (Dietz *et al.*, 2003; Mancini *et al.*, 2017). Annual licensing schemes are a possible option as long as the animal population use is below its 'sustainable tourism yield' (STY), i.e. the maximum use we can make of this population for tourism safely without endangering the sustainability of tourism operations. However, as we have seen above, this STY will vary annually, sometimes widely, depending on the ecological conditions (Lusseau, 2014). Providing annual licenses close to STY means that some will need to be revoked in some years. However, regulatory frameworks rarely provide the foundations to make these unpalatable management decisions. This leads to a ratcheting effect yielding a conservation threat or local economic disaster (Pirotta & Lusseau, 2015).

What is Cryptogovernance?

Blockchains were developed for similar problems in electronic transactions. They act as a ledger system that records the transactions of an asset (the blocks) that may change hands many times over short periods of time and where the identity of the owners need not be clearly known to all in the chain of transactions (Figure 12.1). As shown in the left part of Figure 12.1, in a blockchain, two individuals wanting to engage in a decentralised trusted transaction raise a transaction request to a peer-to-peer (P2P) network. If the transaction meets the criteria of the P2P network (those can vary across ecological, natural and social dimensions), the P2P network carries out computing work to validate the transaction and create a unique token (block) associated to it. This block is likely to be associated to other blocks created by the P2P network and this relationship is lodged when the block is entered in an immutable and permanent electronic ledger, shown on the right of Figure 12.1. As the block is permanent and unique, nothing prevents the block owner from trading it as well. Therefore, blockchain offers the same opportunities to develop derivatives and add value as the stock exchange market did for commodities in the 19th century with future contracts.

This decentralised transactional approach avoids the need for a third party to authorise transactions. It also displaces the issue of trust as individual actors in the chain of transactions need not be trusted because trust is distributed in the blockchain system which jointly ensures the authenticity of the 'seller' and the validity of the currency in the transaction. It also means that property right is clearly assigned at all time, whoever holds the block has the right, and can be retrospectively queried. This is something we cannot do with a centralised currency system where the currency cannot be linked one-to-one to the commodity obtained for its transaction (our coins and bank notes can be used many times for multiple transactions).

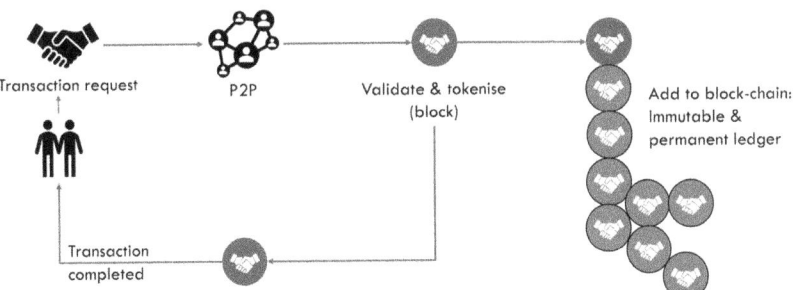

Figure 12.1 How blockchain works: The chain of events when a transaction is requested

Current blockchain projects for environmental sustainability

The blockchain approach can provide advantages to govern common goods (Andoni *et al.*, 2017; Baynham-Herd, 2017; Chapron, 2017; Dao, 2018; Herweijer *et al.*, 2018). The cryptogovernance of common goods relies on using blockchain technology to handle a complete recording of transactions in the way those goods are used. It offers traceability in what were previously thought of as intractable transactions, for example by digitally assigning ownership in intractable situations as is encountered with wildlife tourism. It offers individuals who do not have access to banking the opportunity to engage in P2P transactions in non-traditional ways; providing new financial incentives for many communities that are currently disenfranchised. It provides a means for governance accountability without the need for strong and trusted institutions.

Several projects are already applying cryptogovernance to sustainability challenges. For example, the SunExchange (2019) project (https://thesunexchange.com) achieves the incentivisation of renewable energy development by linking a 'remote' global market to local communities that are 'resource rich' but lack the infrastructure to exploit it. In this context, individuals who want to invest in renewable energy but are located in 'renewable-energy poor' locations can invest in the development of renewable energy infrastructure in 'solar-energy-rich' locations that do not have the scope to invest in the development of such infrastructure and do not attract classical large investors. In this transaction, the small investors receive an equitable return on their investment based on the energy production of the project.

The GainForest project (https://gainforest.org) (Dao, 2018) also aims to achieve incentivisation of environmental sustainability by allowing local farmers to receive returns on forest investment from small investors who value the Amazon forest even though they do not have a stake in its preservation beyond holding its intrinsic global value high. Blockchain here provides a democratisation of investment which through scaling manages to raise substantial funds to allow the value of forest plots to increase without exploiting them. In this way, the microeconomics drivers of land use in the Amazon are changed and farmers neighbouring the native forest can invest in preserving it in a financially sound approach without having to extract goods (timber) from it.

The Japanese Gibier Association is assuring the provenance of game meat delivered to restaurant customers using blockchain (http://forums.mijin.io/en/992.html). Customers are provided with a QR code (barcode) that enables them to completely trace the meat they eat from where it was hunted to the table.

Finally, the Fishcoin project (https://fishcoin.co) aims to do the same for small-scale fisheries. In this project, not only does blockchain technology provide a means to trace the provenance of fish and their path in the supply chain, but it also incentivises providing this data by increasing the worth of the catch with data collection. This is particularly important to manage illegal, unreported and unregulated (IUU) fisheries which often lack infrastructure investment in data-driven management and governance. In this instance, it is the customer-driven demand for traceability that funds data provision, removing the financial burden for data provision from small-scale fishers.

Cryptogovernance of Wildlife Tourism

How might those principles be applied to wildlife tourism? Importantly, how can blockchain technology be applied to improve on the current management hurdles we are facing? Applying blockchain technology to the cryptogovernance of wildlife tourism will rely on the tokenisation of each potential interaction between wildlife and tourists. The following text describes the adoption of blockchain in wildlife tourism where wildlife access rights cannot be allocated in a traditional way (e.g. whale watching), illustrated graphically in Figure 12.2.

As in the generic case illustrated in Figure 12.1, the initial tokenisation is done by a P2P network following a set of rules that are defined *a priori* by co-members of a collective that is set up to manage this form of

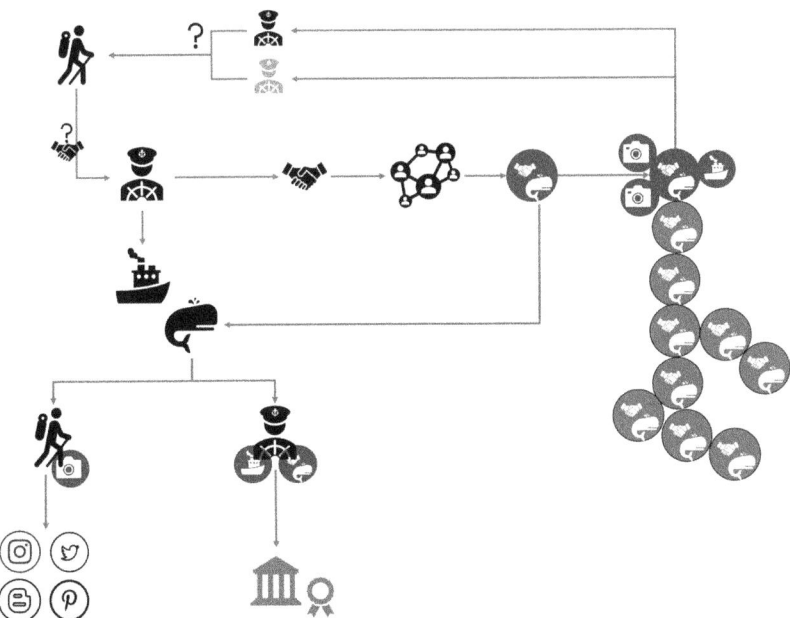

Figure 12.2 Proof-of-concept diagram of how blockchain technology could be applied for the cryptogovernance of wildlife tourism. Icons are templated to Figure 12.1 to demonstrate the link to current blockchain applications

wildlife tourism. In this instance, a tourist makes a choice about a tour operator based on their track record in the blockchain of wildlife interaction 'values'. This is shown in Figure 12.2, at the beginning of the process. Once this choice is made, tourists elicit a transaction request (wildlife tourism engagement) with the operator who then elicits the tokenisation of this transaction. The P2P network automatically assesses the validity of the request depending on the network's agreement for tokenisation. Once the block is created, the interaction can take place and it is lodged in the blockchain ledger.

The tokenisation of the interaction transfers the value of this interaction to the block created. The value of this block can be assessed using both tourist and operator automated information. Blocks can be issued at a relevant time scale, and once the block is used, the value of the encounter is encapsulated in that unique block. Importantly, tourists can contribute to the block value and produce derivatives from it. For example, the tourist contribution can be through the content they generated during that wildlife interaction such as photos, as shown by the cameras on the right of Figure 12.2. As shown at the bottom of Figure 12.2, both tourists and operators can elicit block derivatives (e.g. photos, demonstration of interaction quality) and extract value from those, for example, through social media, as collateral to obtain a loan to invest in infrastructure, or as certification of quality to access privileged interactions (e.g. rarely allowed interactions at a sensitive time or place with wildlife). The derivatives can themselves add value to the original block, creating a virtuous circle between tourists and operators. This incentivisation to collect data about interaction quality also generates the information required for policing. This information can be used to swiftly act on rule-breaking and offer a range of opportunities to automate the consequences of rule-breaking, from tokenisation denial to capital reduction.

Challenges and benefits of cryptogovernance

Block ownership also assigns responsibility for the interaction so that rule adherence is incentivised. Poor-quality interactions will have lower value and interactions without a token issued will have no value (tourists not engaging). In rule-breaking instances, the P2P algorithms allow penalties to be swiftly issued (Ostrom *et al.*, 1999), such as refusal to issue blocks, eviction from the collective or penalty on a membership fee. In this process, the wildlife cost incurred by rule-breaking is valued in the blockchain. Traditionally, rule-breaking incentives increase as the species' abundance declines and the supply:demand ratio therefore decreases. In this new paradigm, those incentives would decrease.

The blockchain itself becomes a valuable tool for tourist decisions when choosing which operator and, indeed, which destination to use. This feedback loop further increases incentives for compliance to collective agreements as defection from those sustainability rules by some operators will decrease the overall value of the blockchain, and its derivatives, for the whole destination.

This collective concept matches the vision of private–public partnerships described in many successful management of commons (Dietz et al., 2003; Mancini et al., 2017; Pirotta & Lusseau, 2015). The tokenisation requires P2P verification of the validity of the information provided for the transaction. This information can take many forms and, indeed, can come from automated sampling such as sampling regimes already used in other wildlife management schemes (e.g. https://earthranger.com, http://www.greatelephantcensus.com, https://oceanmind.global) (Chase et al., 2016; Quattrocchi & Maynou, 2018; Russo et al., 2018). While the validation process has traditionally been associated with the amount of work the P2P had to do for the tokenisation (proof of work [PoW]), faster and more efficient algorithms are now available to achieve the same goal without jeopardising trust in the process by proving a share in the blockchain (proof of stake [PoS]), authority (PoA), history (PoH) or importance (PoI) (Herweijer et al., 2018). These approaches provide a mechanism to treat each transaction, i.e. interaction request, separately in a trusted manner (Adams & Tomko, 2018; De Domenico & Baronchelli, 2019). Therefore, it provides a means to manage and value wildlife use at the relevant scale (one interaction at a time).

Still, some challenges remain. One problem might be that such an approach, as in any classical tragedies of the commons, could encourage operators to quickly maximise the number of interactions they make in order to maximise their share of the wildlife resource. Cryptogovernance can ensure environmental sustainability in such an instance, as the market closes when all 'allowable' interactions have been spent for the year. However, it could increase socioeconomic volatility with, for example, a drive to a very short tourism period, when all annual transactions are spent very quickly, leading at best to monopolisation. The advantage of the P2P network automated management approach is that more complex allocation algorithms can be designed as each transaction is treated independently as opposed to operators being provided with a transaction quota for a given time period. Hence, both annual and daily maximum allocations can be designed. In addition, the co-members of the network can design allocation rules that are not solely based on wildlife maximum sustainable yield of interactions but on broader aspects of STY, including social dimensions. In such approaches, equitable shares can be agreed. Indeed, such sharing algorithms can be incentivised with other blockchains based on the destination's social capital.

This approach also empowers communities in wildlife management. Currently, national institutions are effectively acting as brokers, trading service provision for wildlife. This trade is based on a globally agreed worth of species which sets national targets on biodiversity, for example (Butchart *et al.*, 2010; Lusseau & Mancini, 2019; UN General Assembly, 2015). It can also be based on national targets set to manage the exploitation of particular species (Roman *et al.*, 2013). This centralisation of trade management has two shortcomings. It can over-value wildlife in some communities living next to a charismatic species highly valued by other people living far away from it (e.g. African elephants [Douglas & Alie, 2014]). While charismatic species such as African elephants are globally highly valued, local communities cannot extract the real worth of this species in such instance because they largely cannot access this 'market' (it is remote). Again, other complementary future technological tourism developments (e.g. virtual reality) may be able to remediate this shortcoming. It can also undervalue wildlife. Local communities that have co-evolved with wildlife can be associated with a rich cultural heritage (Wilson, 2002). However, more globally, this wildlife may be considered a pest to be managed to minimise its impact on other human activities; this is the case of wolves (Sjolander-Lindqvist, 2009). Centralised 'trading' through its national, and increasingly international, averaging effect does not allow for communities that are 'wildlife rich' to fully exploit their assets. Decentralising the non-lethal wildlife trade offers an opportunity to let those communities become the trading brokers for their neighbouring wildlife. Doing so also provides a means to restore value to wildlife co-existence and helps to account for intangible local wildlife benefits, which tend to get lost in a national valuation approach, such as the health benefits of wildlife exposure (Bowler *et al.*, 2010; Cox *et al.*, 2017). That is also why community-led public–private wildlife management partnerships seem to be so effective in achieving sustainability both theoretically (Mancini *et al.*, 2017) and in practice (Ostrom *et al.*, 1999). However, one common hurdle to establishing such partnerships is often the lack of sufficiently flexible financial infrastructures, or indeed the lack of trusted financial infrastructures in some countries, to handle their transactions. Blockchain technology solves this problem.

Conclusion: Revaluing Wildlife Use

Estimating the net financial benefits of wildlife tourism remains complex. Its social and environmental net benefits are even harder to comprehend. Currently, it is not clear that this industry extracts a value from wildlife use which matches what we deem this wildlife is worth for the socio-ecological impact it generates (Lusseau & Mancini, 2018; Sandbrook, 2010). One problem is that the ownership of the right to use wildlife is not clearly defined.

This chapter has discussed the possibility of seeing non-lethal wildlife use through a cryptogovernance lens and has argued that this might help to move away from the aforementioned problem of unclear ownership of wildlife. Here, the P2P question becomes whose wildlife use will be impaired by the extraction of usage through tourism. Wildlife will have non-financial value to the communities where it lives. Wildlife will contribute to the social capital of these communities in multiple ways from health improvement to cultural identity. Here, a tokenisation of wildlife use provides a means to value this transaction without having to centralise it through a common denominator, as we do when pricing an environmental asset using a central currency in traditional environmental economics. The intangible value of blocks can be built in the development of derivatives from the block so that wildlife value is no longer linked only to the price associated with its 'first sale'. Like the stock exchange market provided a means to unlock the worth of commodities in the 19th century with the innovation of future contracts, blockchain technology provides a means to unlock the worth of common good usage in the 21st century.

This, of course, assumes that we have learned about opportunities and mistakes emerging from stock market practices in exchanging stocks and developing derivatives on those future contracts (Bessembinder & Seguin, 1992; Clapp & Helleiner, 2012; Szyszka, 2011). For example, the development of agricultural derivatives can introduce new risks for producers and decrease their resilience to change rather than help them weather it (Isakson, 2015). Care must be taken to ensure that we learn from past derivative market failures.

Blockchain technology provides a way to reconcile wildlife worth in a remote market with its value in local communities and enable mechanisms for private–public decentralised partnerships that can navigate wildlife tourism away from the tragedy of the commons. Yet, most communities in dire need of these developments do not have the capacity to implement them. We need to see a concerted effort from inter-governmental organisations, development banks and the crypto sectors to develop the foundation for this new way to connect wildlife-rich communities to the rest of the world.

Acknowledgements

I would like to thank the Greyhope Bay project in Aberdeen (https://greyhopebay.com) and the Aberdeen Science Centre for offering me the platform to discuss these ideas with the public on multiple occasions. These townhall meetings helped me to develop further ideas presented in this chapter. I would also like to thank Giovanna Bertella for comments on the manuscript that substantially improved it.

References

Adams, B. and Tomko, M. (2018) A critical look at cryptogovernance of the real world: Challenges for spatial representation and uncertainty on the blockchain (short paper). *Leibniz International Proceedings in Informatics* 114, 1–6.

Amo, L., López, P. and Martín, J. (2006) Nature-based tourism as a form of predation risk affects body condition and health state of *Podarcis muralis* lizards. *Biological Conservation* 131 (3), 402–409.

Andoni, M., Robu, V. and Flynn, D. (2017) Blockchains: Crypto-control your own energy supply. *Nature* 548 (7666), 158. https://doi.org/10.1038/548158b

Baum, J., Cumming, G.S. and De Vos, A. (2017) Understanding spatial variation in the drivers of nature-based tourism and their influence on the sustainability of private land conservation. *Ecological Economics* 140, 225–234. https://doi.org/10.1016/j.ecolecon.2017.05.005

Baynham-Herd, Z. (2017) Technology: Enlist blockchain to boost conservation. *Nature* 548 (7669), 523. https://doi.org/10.1038/548523c

Baynham-Herd, Z., Amano, T., Sutherland, W.J. and Donald, P.F. (2018) Governance explains variation in national responses to the biodiversity crisis. *Environmental Conservation* 45 (4), 407–418. https://doi.org/10.1017/S037689291700056X

Bessembinder, H. and Seguin, P.J. (1992) Futures: Trading activity and stock price volatility. *The Journal of Finance* 47 (5), 2015–2034. https://doi.org/10.1111/j.1540-6261.1992.tb04695.x

Bowler, D.E., Buyung-Ali, L.M., Knight, T.M. and Pullin, A.S. (2010) A systematic review of evidence for the added benefits to health of exposure to natural environments. *BMC Public Health* 10 (1), 456. https://doi.org/10.1186/1471-2458-10-456

Butchart, S.H.M., Walpole, M., Collen, B., van Strien, A., Scharlemann, J.P.W., Almond, R.E.A., Baillie, J.E. and Bomhard, B. (2010) Global biodiversity: Indicators of recent declines. *Science* 328 (5982), 1164–1168. https://doi.org/10.1126/science.1187512

Chapron, G. (2017) The environment needs cryptogovernance. *Nature* 545 (7655), 403–405. https://doi.org/10.1038/545403a

Chase, M.J., Schlossberg, S., Griffin, C.R., Bouché, P.J.C., Djene, S.W., Elkan, P.W., Ferreira, S., Grossman, F., Kohi, E.M., Landen, K., Omondi, P., Peltier, A., Selier, J.S.A. and Sutcliffe, R. (2016) Continent-wide survey reveals massive decline in African savannah elephants. *PeerJ* 4, e2354. https://doi.org/10.7717/peerj.2354

Christiansen, F. and Lusseau, D. (2015) Linking behavior to vital rates to measure the effects of non-lethal disturbance on wildlife. *Conservation Letters* 8 (6), 424–431. https://doi.org/10.1111/conl.12166

Clapp, J. and Helleiner, E. (2012) Troubled futures? The global food crisis and the politics of agricultural derivatives regulation. *Review of International Political Economy* 19 (2), 181–207. https://doi.org/10.1080/09692290.2010.514528

Cousins, J.A., Sadler, J.P. and Evans, J. (2010) The challenge of regulating private wildlife ranches for conservation in South Africa. *Ecology and Society* 15 (2), 28–48. https://doi.org/10.5751/ES-03349-150228

Cox, D., Shanahan, D., Hudson, H., Fuller, R., Anderson, K., Hancock, S., and Gaston, K.J. (2017) Doses of nearby nature simultaneously associated with multiple health benefits. *International Journal of Environmental Research and Public Health* 14 (2), 172. https://doi.org/10.3390/ijerph14020172

Dao, D. (2018) Decentralized sustainability – David Dao – Medium. See https://medium.com/@daviddao/decentralized-sustainability-9a53223d3001 (accessed 14 June 2019).

De Domenico, M. and Baronchelli, A. (2019) The fragility of decentralised trustless socio-technical systems. *EPJ Data Science* 8 (1), 2. https://doi.org/10.1140/epjds/s13688-018-0180-6

Dietz, T., Ostrom, E. and Stern, P.C. (2003) The struggle to govern commons. *Science* 302 (5652), 1907–1912.

Douglas, L.R. and Alie, K. (2014) High-value natural resources: Linking wildlife conservation to international conflict, insecurity, and development concerns. *Biological Conservation* 171, 270–277. https://doi.org/10.1016/J.BIOCON.2014.01.031

Ellenberg, U., Setiawan, A.N., Cree, A., Houston, D.M. and Seddon, P.J. (2007) Elevated hormonal stress response and reduced reproductive output in Yellow-eyed penguins exposed to unregulated tourism. *General and Comparative Endocrinology* 152 (1), 54–63.

Herweijer, C., Waughray, D. and Warren, S. (2018) *Fourth Industrial Revolution for the Earth Series: Building Block(chain)s for a Better Planet*. Geneva: World Economics Forum. See http://www3.weforum.org/docs/WEF_Building-Blockchains.pdf (accessed 21 January 2019).

Higham, J.E.S., Bejder, L. and Lusseau, D. (2009) An integrated and adaptive management model to address the long-term sustainability of tourist interactions with cetaceans. *Environmental Conservation* 35 (4), 294–302.

Higham, J.E.S., Bejder, L., Allen, S.J., Corkeron, P.J. and Lusseau, D. (2016) Managing whale-watching as a non-lethal consumptive activity. *Journal of Sustainable Tourism* 24 (1), 73–90. https://doi.org/10.1080/09669582.2015.1062020

Isakson, S.R. (2015) Derivatives for development? Small-farmer vulnerability and the financialization of climate risk management. *Journal of Agrarian Change* 15 (4), 569–580. https://doi.org/10.1111/joac.12124

Kinnaird, M.F. and O'Brien, T.G. (2012) Effects of private-land use, livestock management, and human tolerance on diversity, distribution, and abundance of large African mammals. *Conservation Biology* 26 (6), 1026–1039. https://doi.org/10.1111/j.1523-1739.2012.01942.x

Lusseau, D. (2005) Residency pattern of bottlenose dolphins *Tursiops* spp. in Milford Sound, New Zealand, is related to boat traffic. *Marine Ecology Progress Series* 295, 265–272.

Lusseau, D. (2014) *Ecological Constraints and the Propensity for Population Consequences of Whale-Watching Disturbances*. Chicago, IL: Cambridge University Press. https://doi.org/10.1017/CBO9781139018166.019

Lusseau, D. and Higham, J.E.S. (2004) Managing the impacts of dolphin-based tourism through the definition of critical habitats: The case of bottlenose dolphins (*Tursiops* spp.) in Doubtful Sound, New Zealand. *Tourism Management* 25 (6). https://doi.org/10.1016/j.tourman.2003.08.012

Lusseau, D. and Mancini, F. (2018) A global assessment of tourism and recreation Conservation threats to prioritise interventions. *Current Biology*. See http://arxiv.org/abs/1808.08399.

Lusseau, D. and Mancini, F. (2019) Income-based variation in sustainable development goal interaction networks. *Nature Sustainability* 2 (3), 242–247. https://doi.org/10.1038/s41893-019-0231-4

Lusseau, D., Slooten, E. and Currey, R.J.C. (2006) Unsustainable dolphin-watching tourism in Fiordland, New Zealand. *Tourism in Marine Environments* 3 (2), 173–178. https://doi.org/10.3727/154427306779435184

MacMillan, D.C. and Phillip, S. (2008) Consumptive and non-consumptive values of wild mammals in Britain. *Mammal Review* 38 (2–3), 189–204. https://doi.org/10.1111/j.1365-2907.2008.00124.x

Mancini, F., Coghill, G.M. and Lusseau, D. (2017) Using qualitative models to define sustainable management for the commons in data poor conditions. *Environmental Science and Policy* 67, 52–60. https://doi.org/10.1016/j.envsci.2016.11.002

Meletis, Z.A. and Campbell, L.M. (2007) Call it consumption! Re-conceptualizing ecotourism as consumption and consumptive. *Geography Compass* 1 (4), 850–870. https://doi.org/10.1111/j.1749-8198.2007.00048.x

Ngoprasert, D., Lynam, A.J. and Gale, G.A. (2017) Effects of temporary closure of a national park on leopard movement and behaviour in tropical Asia. *Mammalian Biology* 82, 65–73. https://doi.org/10.1016/J.MAMBIO.2016.11.004

Ostrom, E. (2009) A general framework for analyzing sustainability of social-ecological systems. *Science* 325 (5939), 419–422.

Ostrom, E., Burger, J., Field, C.B., Norgaard, R.B. and Policansky, D. (1999) Revisiting the commons: Local lessons, global challenges. *Science* 284 (5412), 278–282.

Pirotta, E. and Lusseau, D. (2015) Managing the wildlife tourism commons. *Ecological Applications* 25 (3), 729–741. https://doi.org/10.1890/14-0986.1.sm

Pirotta, E., Harwood, J., Thompson, P.M., New, L., Cheney, B., Arso, M., Hammond, P.S., Donovan, C.R. and Lusseau, D. (2015) Predicting the effects of human developments on individual dolphins to understand potential long-term population consequences. *Proceedings of the Royal Society B: Biological Sciences* 282 (1818). https://doi.org/10.1098/rspb.2015.2109

Pirotta, E., Booth, C.G., Costa, D.P., Fleishman, E., Kraus, S.D., Lusseau, D., Moretti, D., New, L.F., Schick, R.S., Schwarz, L.K., Simmons, S.E., Thomas, L., Tyack, P.L., Weise, M.J., Wells, R.S. and Harwood, J. (2018) Understanding the population consequences of disturbance. *Ecology and Evolution* 8 (19), 9934–9946. https://doi.org/10.1002/ece3.4458

Quattrocchi, F. and Maynou, F. (2018) Spatial structures and temporal patterns of purse seine fishing effort in the NW Mediterranean Sea estimated using VMS data. *Fisheries Management and Ecology* 25 (6), 501–511. https://doi.org/10.1111/fme.12325

Reed, S.E. and Merenlender, A.M. (2008) Quiet, non-consumptive recreation reduces protected area effectiveness. *Conservation Letters* 1 (3), 146–154. https://doi.org/10.1111/j.1755-263X.2008.00019.x

Roman, J., Altman, I., Dunphy-Daly, M.M., Campbell, C., Jasny, M. and Read, A.J. (2013) The Marine Mammal Protection Act at 40: Status, recovery, and future of U.S. marine mammals. *Annals of the New York Academy of Sciences* 1286 (1), 29–49. https://doi.org/10.1111/nyas.12040

Russo, T., Morello, E.B., Parisi, A., Scarcella, G., Angelini, S., Labanchi, L., Martinelli, M., D'Andrea, L., Santojanni, A., Arneri, E. and Cataudella, S. (2018) A model combining landings and VMS data to estimate landings by fishing ground and harbor. *Fisheries Research* 199, 218–230. https://doi.org/10.1016/J.FISHRES.2017.11.002

Sandbrook, C.G. (2010) Local economic impact of different forms of nature-based tourism. *Conservation Letters* 3 (1), 21–28. https://doi.org/10.1111/j.1755-263X.2009.00085.x

Sjolander-Lindqvist, A. (2009) Social-natural landscape reorganised: Swedish forest-edge farmers and wolf recovery. *Conservation and Society* 7 (2), 130. https://doi.org/10.4103/0972-4923.58644

SunExchange (2019) Upcoming Sun Exchange Solar Project: Senior citizens go solar. See https://medium.com/@alias_73214/upcoming-sun-exchange-solar-project-senior-citizens-go-solar-890d83888dc2 (accessed 15 June 2019).

Szyszka, A. (2011) The genesis of the 2008 global financial crisis and challenges to the neoclassical paradigm of finance. *Global Finance Journal* 22 (3), 211–216. https://doi.org/10.1016/J.GFJ.2011.10.011

Twining-Ward, L., Li, W., Bhammar, H. and Wright, E. (2018) Supporting sustainable livelihoods through wildlife tourism. See https://openknowledge.worldbank.org/handle/10986/29417 (accessed 21 Jaunaruy 2019).

Tyne, J.A., Johnston, D.W., Rankin, R., Loneragan, N.R. and Bejder, L. (2015) The importance of spinner dolphin (*Stenella longirostris*) resting habitat: Implications for management. *Journal of Applied Ecology* 52 (3), 621–630. https://doi.org/10.1111/1365-2664.12434

UN General Assembly (2015) *Transforming Our World: The 2030 Agenda for Sustainable Development*. New York: United Nations. See http://www.un.org/ga/search/view_doc.asp?symbol=A/RES/70/1&Lang=E (accessed 29 June 2017).

Wilson, P.I. (2002) Native peoples and the management of natural resources in the Pacific Northwest: A comparative assessment. *American Review of Canadian Studies* 32 (3), 397–414. https://doi.org/10.1080/02722010209481668

World Tourism Organization (2015) *Towards Measuring the Economic Value of Wildlife Watching Tourism in Africa*. Madrid: UNWTO. See https://sustainabledevelopment.un.org/content/documents/1882unwtowildlifepaper.pdf (accessed 7 April 2018).

13 Final Reflections: Travel Notes, Postcards, Treasures and Dragons

Giovanna Bertella

This book has presented several future scenarios concerning wildlife tourism, including different settings, animal species and encounters. The described scenarios derive from the authors' reflections on trends, their imagination and from the adoption of various theoretical perspectives and different focuses. The futures depicted in such scenarios make us reflect on what animals and encounters the tourists will perceive as wild and extraordinary, and what values wildlife tourism might build on and promote. In the first chapter, it was proposed that this book could be viewed as a journey to *terra incognita*, a sort of exploration of the unknown, the future of wildlife tourism. Along our journey, in the 11 chapters of Parts 1, 2 and 3, *terra incognita* has emerged in all its complexities and contractions. It is now time to ask ourselves what we have experienced from this journey. This will be the final contribution of this book: the identification of some lessons about what the future of wildlife tourism might look like and how to proceed in its development and implementation.

We can imagine being at the end of the exploration of *terra incognita*, almost ready to pack our things and travel back to the present. Before leaving, we can send a couple of postcards reporting our impressions and experiences about *terra incognita*. We can then collect our memories from this long journey and reflect on the scenarios we have been presented with, and about the possible treasures guarded by the dragons of *terra incognita*.

Travel Notes and a Postcard from Part 1: Paths Towards the Futures of Wildlife Tourism

The journey started with Qingming Cui. Among the various inspirational sources for this first part of our journey are the works by sociologist John Urry (2013, 2015, 2016). Qingming Cui has unfolded before us three futures deriving from the possible effects of climate change.

The first futuristic scenario we have learned about from Qingming Cui has taken us to a sustainable future, characterised by green marketing and technological innovations. The second scenario has claimed the need to reduce social production and consumption to attain a sustainable living environment. The third scenario that is considered by the author, the most possible global future, has taken us to a dark future. We have seen that, in this future, most of the remaining wildlife inhabit militarised nature reserves. Only a very limited group of wealthy and powerful tourists has the opportunity to experience wildlife tourism in nature reserves, while the vast majority of the global population is poor and view wild animals mainly as a source of food. Qingming Cui has made us reflect on the fact that the sustainability of the tourism industry in response to climate change needs a macro social transformation rather than micro changes within the industry.

Our journey along the paths to the futures of wildlife tourism continued with Rie Usui and Carolin Funck. The scenarios they have created suggest to us the possibility that sustainability, which is often reported as the main approach to be used when thinking about and directing our actions towards the future, might be challenged by other approaches. The off the beaten path that the authors have described to us relies on ecofeminism (Kheel, 2008; Plumwood, 1993; Vance, 1997; Warren, 1990). Rie Usui and Carolin Funck have invited us to reflect deeply on the anthropocentric approach of sustainability and, in particular, on the view of nature, including wildlife, as a resource for humanity. Ecofeminism proposes to go beyond the duality of humans and wild animals. Further reflections on such a dualistic way of thinking have brought the authors to realise that the case of domesticated animals living in semi-wild conditions, as the rabbits of Ōkunoshima Island, is often overlooked in tourism. The authors have suggested that this might be because of their unsettling position between wild and domestic, and their ability to transgress such a boundary. In such cases, the responsibility of humans, and in particular the tourism sector, is very evident, and the adoption of ecofeminism is fruitful in highlighting the values of care and the related practical challenges.

Georgette Leah Burns and Judith Benz-Schwarzburg were our guides in the last part of this initial journey. They have invited us to observe the ethical landscape of wildlife tourism (Cataldi, 2002; De Grazia, 2005; Foucault, 1994; Rowlands, 2012). They have noted that considerable changes in our relationships with animals are occurring and will continue in the future. In their futures, the authors have imagined powerful and immersive wildlife experiences for tourists. The authors have warned us about a possible dark side to these experiences. Georgette Leah Burns and Judith Benz-Schwarzburg have suggested that the design of wildlife experiences, now and in the future, should be based on ethical care and consideration that include the preservation of animal dignity.

Thus, what could we write on our postcard from this first part of our travel to *terra incognita*? What have we experienced and learned about the possible paths to the futures? Maybe our postcard might read as follows:

> The paths to *terra incognita* are various and have been illustrated to us by five exceptionally engaged scholars who have acted as guides in our attempt to learn about possible futures. While pondering which path to choose, we should make sure to have a broad overview about our options, what is influencing and might affect these paths in the future. We should reflect deeply and clarify our ethical position in relation to the natural environment, the wildlife and also other animals. We should reflect on which are the values and disvalues attached to this position. There might not be an absolute best path, and considering and debating the different choices might be what makes this journey valuable.

Travel Notes and a Postcard from Part 2: Human–Animal Encounters

As promised in the first chapter of this book, in the second part of our journey, we had the opportunity to witness, experience and reflect on various human–animal encounters.

Jessica Bell Rizzolo was the first author to accompany us on this part of the journey. The three futures depicted by Jessica Bell Rizzolo derive from the complicated mechanism within our ethical position towards animals, power relations, sociocultural trends, commercial interests and digital egocentrism (Fennell, 2012; Iqani & Schroeder, 2016; Lim, 2016; Mutalib, 2018; Nekaris *et al.*, 2013). In her chapter, Jessica Bell Rizzolo has warned us about the increasing popularity of selfies with wildlife and the amplified effects that such practice can have due to the use of social media. She has pointed out the severe and persistent effects of selfie safaris on wildlife, and illustrated how decisions by tourists and the social media companies can dramatically alter the future of wildlife on both an individual and a species level.

Ronda J. Green invited us to visit a futuristic zoo developed on the ideal balance between the needs and desires of the animals and the tourists. She has highlighted the need for more knowledge and research that can help captive facilities truly understand the animals. What we might need, she has suggested, is to improve the captive animals' life, not only limiting possible negative effects but also introducing elements that can enrich their everyday life (Dawkins, 2004; Reading *et al.*, 2013). Ronda J. Green has not limited her reflections to animal welfare and basic needs but has included the important element of life enjoyment for the animals. We have joined Ronda J. Green in her fictive utopian zoo, and seen her entertaining a lemur and a parrot. In her futuristic zoo, the role of spectator and entertainer might be inverted: the zoo visitor, with his/her presence, might be a source of entertainment for the animals.

In the journey with Gianna Moscardo, we visited Northern Australia, joining a tour with our guide Bob, together with Freeman Tilden, the father of heritage interpretation. The tour has been presented in two versions. In a version that is more in line with the concept of interpretation, new technologies have triggered our curiosity and engaged us in a mindful way. However, Gianna Moscardo has warned us that technologies themselves might not be the main element of these high-quality, engaging and responsible tourist–animal encounters. She has made us notice that the main difference between a utopian and dystopian scenario about wildlife encounters is not the type or number of new technologies used but the underlying value placed on the wildlife. This reflection echoes some of the lessons learned in the first part of our journey. In line with the concluding reflections by Georgette Leah Burns and Judith Benz-Schwarzburg, Gianna Moscardo has concluded this part of our journey by suggesting that greater attention needs to be paid to animal ethics.

This second part of the journey closes with Giovanna Bertella, who, reflecting on human–animal relations (Derrida & Wills, 2002; Gaard, 1993; Gruen, 2015), has focused her attention on what the animals involved in tourist activities might think about such involvement, the humans they meet and, more generally, humanity. Giovanna Bertella has imagined a future where orca–human communication is possible. Then, she has presented three alternative scenarios concerning the discussions among scientists and investors about the future of orca tourism in Northern Norway. In two scenarios, orca tourism has been presented as possible and characterised by different mutual benefits. Giovanna Bertella has invited us to reflect on what this might require from the tourism sector, the tourists, as well as governmental agencies. In the third future scenario created for us by Giovanna Bertella, the animals have clearly stated that they are not willing to be involved in any tourism. With such an uncomfortable possibility in our mind, the author has left us with a big question to ponder, namely about wildlife tourism encounters as coercive meetings between humans and wild animals.

The postcard we might send from this second part of our journey in *terra incognita* might include the following reflections:

> In *terra incognita* there are various types of tourist–animal encounters. Considering our relations with wildlife, a shift seems to be occurring. Although the human dimension of these encounters is still considered important due to its emotional and educational potentials, the focus of the four authors who have been the imaginative guides of our travel seems to be on the animals. Once considered mere attractions, the animals, thanks to scientific advancements and the engagement of caring people, might be able to make their voices heard in the future. Will the

tourism sector, the tourists and the relevant governmental agencies be willing to listen?

Travel Notes and a Postcard from Part 3: Technology Advancements

Hindertje Hoarau-Heemstra and Anne-Mette Hjalager were the first authors we met at the beginning of our last exploration of *terra incognita*. They have introduced us to Zoë, a tourist. Together with Zoë, we have joined a futuristic whale watching tour. On board an eco-friendly vessel, we have had the opportunity to hear and understand some whale conversations. We have been informed about the business practices and values of the whale watching company. We have dreamed about a future where respectful and engaging wildlife experiences occur thanks to collaborative eco-innovation processes (e.g. De Marchi, 2012; De Medeiros *et al.*, 2014; Divisekera & Nguyen, 2018). Hindertje Hoarau-Heemstra and Anne-Mette Hjalager have explained to us that the businesses of the future will need to extend their boundaries and develop collaboration with stakeholders, such as knowledge intermediaries and technology suppliers.

Stepping down from the whale watching boat, we met Mikko Äijälä, Titta Jylkäs, Vésaal Rajab, Tytti Vuorikari and, not least, Alanis, a dog that we will hardly forget. We have been mushing in the Arctic, and experienced the wild nature from a new perspective. Our human guides have explained to us that, once we have recognised the agency of animals, wilderness experiences will be extremely engaging and memorable. To create this scenario, Mikko Äijälä, Titta Jylkäs, Vésaal Rajab and Tytti Vuorikari have adopted service and speculative design to provide us with future-oriented tools (Dunne & Raby, 2013; Miettinen & Koivisto, 2009; Sanders & Stappers, 2008). In this future, modern technology has served as a means for mutual understanding between humans and animals and therefore contributed to the recognition of animal agency.

The journey continued with Daniel William Mackenzie Wright, who showed us a leaflet from a zoo and aquatic cloning centre. We have considered all the services offered by this centre, and found ourselves wondering whether or not we wanted to visit it. The future scenario imagined by Daniel William Mackenzie Wright has provoked many thoughts: we have been confronted with difficult questions regarding our attitudes towards animals and animal cloning, and our motivations and responsibilities as tourists. Exploring the future with Daniel William Mackenzie Wright has provided us with one more opportunity to think about our ethical position towards animals, and what type of captive wildlife institutions we might want for the future (Kagan *et al.*, 2018; Safina, 2018; Zimmermann *et al.*, 2007).

David Lusseau is the guide we met at the end of our long journey in *terra incognita*. The journey finishes in a similar way to how it had begun, thinking about sustainability. We have explored a future where technology can help us to estimate the value of wildlife and avoid the tragedy of the commons of resources with no clear ownership. David Lusseau has explained to us the mechanism of blockchains and cryptogovernance (Andoni et al., 2017; Baynham-Herd, 2017; Chapron, 2017). He has also warned us that not all communities might have the capacity to implement such technology, and collaboration among such communities and intergovernmental organisations will be crucially important.

At the end of this third part of our journey in *terra incognita*, we can now write our final postcard:

> Technology and collaboration have emerged as important factors to innovate wildlife tourism. Through technological devices, we have better understood what the wild animals might care about and feel. We have learned about the possibility of finding technological solutions to the challenges of wildlife tourism, using the experiences gained from other sectors and through collaboration. In our journey, we have experienced only a few ways through which technology will shape wildlife tourism and we are sure that in *terra incognita* many other technological solutions might be found.

Treasures and Dragons in *Terra Incognita*

The 17 authors of this book, the guides of our journey to *terra incognita*, have described the future of wildlife tourism as populated by many different animals and, not least, humans characterised by various values and perspectives. Plurality and diversity have characterised our journey. It is here advanced that this is the treasure we have discovered along our journey to *terra incognita*. Exploring the future is questioning our present, our beliefs and values. In this sense, scenarios development is a critical and fruitful form of scientific inquiry (Amer et al., 2013; Burnam-Fink, 2015; Fahey & Randal, 1998; Lee, 2012; Nassauer & Corry, 2004; Varum et al., 2011; Yeoman, 2012, 2019).

Based on the identification of various change drivers and their critical analysis, the authors created several scenarios, some utopian, responding to the authors' hopes, and others dystopian, as a result of the authors' fears and doubts (Inayatullah, 2008; Robertson & Yeoman, 2014; Schoemaker, 1993; van Notten et al., 2003; Yeoman & Postma, 2014). The authors have used various sources of inspiration: in addition to the scientific literature and reports on tourism, concepts relevant to environmental and animal ethics have also been discussed, together with elements deriving from real events, existing companies and fiction (Bina et al., 2017; Dunne & Raby, 2013; Parry & Johnson, 2007; Richardson & St. Pierre, 2005; Yeoman & Mars, 2012). The result is a representation of the future as a dynamic mosaic, with some bright and many dark areas.

The epigraph of the first chapter reads: *Terra Incognita: hic sunt dracones* (Unknown Land: here are dragons). This has been related to the possibility for the readers to encounter dragons while exploring the future of wildlife and the possible existence of some sort of treasure. It is now the time to ask: What about the dragons guarding the treasure? It might be advanced that, metaphorically, the dragons are the obstacles to a varied, creative, critical and responsible way to think about the future. The sometimes sharp contrast between the numerous scenarios presented in this book highlights the necessity to face these dragons together, giving voice to different perspectives and values and, not least, to the animals involved in tourism. Alternatively, the dragons of *terra incognita* can be understood as the animals themselves. Fierce and fearless as only wild animals can be, they defend their treasure, which is the possibility to have a future.

References

Amer, M., Daim, T.U. and Jetter, A. (2013) A review of scenario planning. *Futures* 46, 23–40.
Andoni, M., Robu, V. and Flynn, D. (2017) Blockchains: Crypto-control your own energy supply. *Nature* 548 (7666), 158.
Baynham-Herd, Z. (2017) Technology: Enlist blockchain to boost conservation. *Nature* 548 (7669), 523.
Bina, O., Mateus, S., Pereira, L. and Caffa, A. (2017) The future imagined: Exploring fiction as a means of reflecting on today's Grand Societal Challenges and tomorrow's options. *Futures* 86, 166–184.
Burnam-Fink, M. (2015) Creating narrative scenarios. *Futures* 70, 48–55.
Cataldi, S.L. (2002) Animals and the concept of dignity. Critical reflections on a circus performance. *Ethics and the Environment* 7 (2), 104–126.
Chapron, G. (2017) The environment needs cryptogovernance. *Nature* 545 (7655), 403–405.
Dawkins, M.S. (2004) Using behaviour to assess animal welfare. *Animal Welfare* 13, S3–7.
De Grazia, D. (2005) Regarding the last frontier of bigotry. *Logos* 4 (2).
De Marchi, V. (2012) Environmental innovation and R&D cooperation. *Research Policy* 41 (3), 614–623.
De Medeiros, J.F., Ribeiro, J.L.D. and Cortimiglia, M.N. (2014) Success factors for environmentally sustainable product innovation: A systematic literature review. *Journal of Cleaner Production* 65, 76–86.
Derrida, J. and Wills, D. (2002) The animal that therefore I am (more to follow). *Critical Inquiry* 28 (2), 369–418.
Divisekera, S. and Nguyen, V.K. (2018) Determinants of innovation in tourism evidence from Australia. *Tourism Management* 67, 157–167.
Dunne, A. and Raby, F. (2013) *Speculative Everything: Design, Fiction, and Social Dreaming*. Cambridge, MA: MIT Press.
Fahey, L.R. and Randal, R. (1998) What is scenario learning. In L. Fahey and R. Randall (eds) *Learning From the Future: Competitive Foresight Scenarios* (pp. 3–43). New York: Wiley.
Fennell, D.A. (2012) *Tourism and Animal Ethics*. London: Routledge.
Foucault, M. ([1975] 1994) *Überwachen und Strafen. Die Geburt des Gefängnisses*. Frankfurt am Main: Suhrkamp.
Gaard, G. (1993) *Ecofeminism: Women, Animals, Nature*. Philadelphia, PA: Temple University Press.

Gruen, L. (2015) *Entangled Empathy: An Alternative Ethic for Our Relationships with Animals*. Brooklyn, NY: Lantern Books.
Inayatullah, S. (2008) Six pillars: Futures thinking for transforming. *Foresight* 10 (1), 4–21.
Iqani, M. and Schroeder, J.E. (2016) # selfie: Digital self-portraits as commodity form and consumption practice. *Consumption Markets & Culture* 19 (5), 405–415.
Kagan, R., Allard, S. and Carter, S. (2018) What is the future for zoos and aquariums? *Journal of Applied Animal Welfare Science* 21 (1), 59–70.
Kheel, M. (2008) *Nature Ethics: An Ecofeminist Perspective*. Lanham, MD: Rowman & Littlefield Publishers.
Lee, M. (2012) *Knowing Our Future: The Startling Case for Futurology*. Oxford: Infinite Ideas Limited.
Lim, W.M. (2016) Understanding the selfie phenomenon: Current insights and future research directions. *European Journal of Marketing* 50 (9/10), 1773–1788.
Miettinen, S. and Koivisto, M. (eds) (2009) *Designing Services with Innovative Methods*. Keuruu: Kuopio Academy of Design.
Mutalib, A. (2018) The photo frenzy phenomenon: How a single snap can affect wildlife populations. *Biodiversity* 19 (3–4), 237–239.
Nassauer, J.I. and Corry, R.C. (2004) Using normative scenarios in landscape ecology. *Landscape Ecology* 19 (4), 343–356.
Nekaris, K.A.I., Campbell, N., Coggins, T.G., Rode, E.J. and Nijman, V. (2013) Tickled to death: Analysing public perceptions of 'cute' videos of threatened species (slow lorises–*Nycticebus* spp.) on Web 2.0 sites. *PloS One* 8 (7), e69215.
Parry, D.C. and Johnson, C.W. (2007) Contextualizing leisure research to encompass complexity in lived leisure experience: The need for creative analytic practice. *Leisure Sciences* 29 (2), 119–130.
Plumwood, V. (1993) *Feminism and the Mastery of Nature*. London: Routledge.
Reading, R.P., Miller, B. and Shepherdson, D. (2013) The value of enrichment to reintroduction success. *Zoo Biology* 9999, 1–10.
Richardson, L. and St. Pierre, E.A. (2005) Writing: A method of inquiry. In N.K. Denzin and Y.S. Lincoln (eds) *The Sage Handbook of Qualitative Inquiry* (pp. 959–978). Thousand Oaks, CA: Sage Publications.
Robertson, M. and Yeoman, I. (2014) Signals and signposts of the future: Literary festival consumption in 2050. *Tourism Recreation Research* 39 (3), 321–342.
Rowlands, M. (2012) *Can Animals Be Moral?* New York: Oxford University Press.
Safina, C. (2018) Where are zoos going – or are they gone? *Journal of Applied Animal Welfare Science* 21 (1), 4–11.
Sanders, E.B.N. and Stappers, P.J. (2008) Co-creation and the new landscapes of design. *Co-design* 4 (1), 5–18.
Schoemaker, P.J. (1993) Multiple scenario development: Its conceptual and behavioral foundation. *Strategic Management Journal* 14 (3), 193–213.
Urry, J. (2013) *Societies Beyond Oil: Oil Dregs and Social Futures*. London/New York: Zed Books Ltd.
Urry, J. (2015) Climate change and society. In J. Michie and C.L. Cooper (eds) *Why the Social Sciences Matter* (pp. 45–59). London: Palgrave Macmillan.
Urry, J. (2016) *What is the Future?* Cambridge: Polity Press.
Vance, L. (1997) Ecofeminism and wilderness. *NWSA Journal* 9 (3), 60–76.
van Notten, P.W.F., Rotmans, J., van Asselt, M.B.A. and Rothman, D.S. (2003) An updated scenario typology. *Futures* 35, 423–443.
Varum, A., Melo, C., Alvarenga, A. and Soeiro de Carvalho, P. (2011) Scenarios and possible futures for hospitality and tourism. *Foresight* 13 (1), 19–35.
Warren, K.J. (1990) The power and the promise of ecological feminism. *Environmental Ethics* 12 (2), 125–146.
Yeoman, I. (2012) *2050 – Tomorrow's Tourism*. Bristol: Channel View Publications.

Yeoman, I. (2019) Does the future have a recipe? *Journal of Tourism Futures* 5 (1), 3–4.
Yeoman, I. and Mars, M. (2012) Robots, men and sex tourism. *Futures* 44 (4), 365–371.
Yeoman, I. and Postma, A. (2014) Developing an ontological framework for tourism futures. *Tourism Recreation Research* 39 (3), 299–304.
Zimmermann, A., Hatchwell, M., Dickie, L.A. and West, C. (2007) *Zoos in the 21st Century: Catalysts for Conservation?* Cambridge: Cambridge University Press.

Index

academia/ academic, 17, 101, 115, 146
activities
    grassroots, 28
    self-organised volunteer, 28
agency, 5, 9, 40, 126–127, 130–31, 134, 136–37, 139, 171
alien species, 27, 33–36
animal
    culture, 100
    encounters, 4, 37, 51, 55, 58–110, 126–28, 136–37, 139, 169–70
    entertainer(-s)/ performer(-s), 46–48
animality, 100, 137–38
Anthropocene, 10, 22, 53
anthropocentric/anthropocentrism 34, 48 49, 138,168
anthropomorphism, 48
aquarium/aquaria, 5, 9, 15, 17, 21, 44, 47, 57, 72, 78, 98–99, 140–45, 101, 110, 147, 150–53
artificial animals, 48–49
artificiality, 47
Attenborough, 146
attitudes, 10, 13, 34, 44, 62, 90, 92, 141, 145–46, 171
augmented reality (AR), 48, 91–93, 120, 134
authenticity, 109, 157

biodiversity, 5, 10, 13–14, 19–20, 35, 68, 114, 147, 150, 162
biophilia, 153, 145

cameras, 15, 78–79, 93, 133, 160
captivity/ captive, 1, 4, 30, 41–42, 44, 49, 51, 54, 57, 62, 67–68, 72–73, 75–76, 82–83, 98, 100, 102, 105, 107–108, 140, 142, 144–45, 150–151, 169, 171
care, 32, 34, 37, 43, 48, 99, 104–5, 143–44, 163, 168, 172
celebrity/ celebrities, 63, 65, 67
certification, 93, 160
climate change, 3, 5, 9–11, 13, 18–20, 22–23, 117, 119, 125, 146, 151, 168, 174
clone/cloning, 2, 5–6, 94, 140–41, 147–50, 153, 171
code of ethics, 40
co-creation, 109, 110, 131, 136–137, 138– 139, 174
co-design, 139, 174
collaboration, 5, 101, 104–6, 113, 115–18, 121, 123, 131, 143–44, 171–72
commercialisation, 44, 52, 92
commodities, 21, 63, 157, 163
community/ communities, 9–10, 16 25, 31, 35, 63, 101, 113–114, 123, 143, 146, 154, 156, 158, 161–63, 172
conflict(-s), 12, 18–19, 29, 42, 48, 107, 129, 144
conservation, 3–5, 14, 19, 37–38, 41–45, 47–48, 50–54, 61–62, 64, 68–72, 77–78, 81–82, 100, 109–10, 138, 140, 142, 149–53, 164
    environmental, 16, 164–65
consumption, 1, 11–12, 16–18, 20, 44–45, 49, 69–70, 146–47, 149–50, 165, 168, 174
cooperation, 122–23
cruelty (animal), 65, 67

cryptogovernance, 157–58, 160–61,
163–64, 172–73
curiosity, 13–14, 85, 107, 142, 170

death, 21, 32, 85, 115
deforestation, 6, 66, 146
degradation, 11, 14, 19
degrowth, 3, 11–13, 16–18 20–23
Derrida, 99, 173
design, 50, 79, 122, 127–28, 131–133,
137–39, 168, 171, 173–74
destinations, 17, 90–91, 99, 154, 156,
160–61
digital experiences, 15–16
dignity, 99, 168
dilemma, 149, 154
Disneyisation, 44–46, 50–51
disruptions, 6, 61, 88, 91
disturbances, 154–55, 166
domination, 32, 34
dystopia/ dystopian, 4, 58, 65–66, 92,
95, 170, 172

ecocentric/ ecocentrism, 3–4, 41–42, 45
eco-certification, 93
ecofeminism, 3–4, 24, 32–36, 38–39,
109, 168, 173–74
eco-innovation(-s), 114–116, 119–123,
171
ecology, 31, 40, 76, 83, 91, 129, 164, 166
economy, 5, 12, 14, 18, 91
ecosystem, 31–32, 34, 114
ecotourism, viii, 36–37, 53, 96, 109–10,
138
education/ educational, 15, 17, 29–30,
41–42, 51, 53, 77–79, 82, 137–138,
140, 142, 149–50, 152
Elysium, 13, 18
emotions, 32, 75, 101, 133–34, 136–37,
144
empathy, 32, 145
endangered species, 10, 19, 21, 61, 66,
68, 72, 79, 83, 143, 152
engagement, 4, 127, 130, 136, 170
enrichment, 61, 76–77, 83, 174
entertainment, vii, 29–31, 43–45, 47–53,
57, 60, 77, 79–80, 142, 152, 170
environmental
awareness, 118
conservation, 16

degradation, 11
decline, 66
economics, 163
ethics, 14, 43, 119, 172
ideologies, 13
impact(-s), 4, 16, 31, 57, 61, 64–65,
118, 120
pollution, 16
sustainability, 158, 161
ethical/ ethics, vii, 2–4, 6, 14–15, 29, 34,
36–38, 40–43, 47–49, 50, 53–54,
83, 100, 108–109, 119, 125, 127,
144, 150,169, 171, 173–174
expectation(-s), 44–45, 51, 87, 92, 123,
127–128, 145
extinction/ extinct species, 1, 5, 10,
13–14, 15, 19, 64–66, 118–119,
140, 142, 146–147, 149

feral, 24, 72, 79
fiction/fictional, 3, 5, 13, 85, 93, 99, 102,
128, 132, 134, 136, 172,–4, 41–42,
45
fishing, 9, 57,117
food, 1, 12, 18–20, 27–33, 59, 73, 75–78,
118, 140, 146–147,
149–150, 168

Gandhi, 140
gaze (animal), 108
gaze (tourist), 49
governance, 114, 117, 156, 158–159
government(-s), 16, 19, 61–62, 66–67,
103, 108, 117, 120
guide, 29, 43, 85, 89, 93, 133, 170–171
guideline(-s), 29, 72–75
green technology, 103

hologram(-s), 15, 79, 118, 120
Hugo, 115
hunt/hunting, 19, 140, 154–155

ICW, 113
innovation, 3, 14–15, 114–116, 118–123,
163, 171
interpretation, 3–4, 15, 85–95, 118,
136–137, 170
Interstellar, 13, 16

job opportunities, 16, 108

Index

knowledge, 9, 13, 30, 42–43, 72–73, 76, 81, 114–116, 121–123, 134, 137, 141–142, 169, 171

landscape, 24, 29, 57, 141, 168,
learning, 77–78, 87, 89–90, 100, 116, 118–119, 121–122
leisure, 17, 19–20, 142

management, 1–4, 14, 27–28, 35–36, 43, 58, 68, 87–88, 90–91, 95, 99, 101, 115–116, 154–156, 159, 161–162
multisensory, 88–90, 134

NGO(-s), 50, 101, 117, 121,
network(-s), 11, 91, 116–117, 121–123, 157, 159–161
non-consumptive, 9–10, 13–15, 17, 19–21, 57–58, 68, 126, 154, 156

outdoor(-s), 126, 127

park(-s), 15, 17, 32, 44, 48, 51, 65, 72,
PETA, 102, 144
pets, 24, 61, 149
personhood (animal), 117–118
petting/petted, 1, 33–35, 42, 48, 50
philosophy/ philosophical, 5, 32, 99
photography, 57,63, 68
policy, 11, 16–17, 29, 100, 114, 147
principles, 4, 40–41, 45, 51, 90, 131, 159
profit/ profitable, 49, 71, 108, 113, 122
protection, 18–19, 27, 35, 40, 47, 49, 94, 113, 117, 119, 144–145

quality, 9, 20, 42, 50, 103, 127, 136, 145, 156, 160, 170

recreation/ recreational, 15, 17, 25–26, 33, 57, 142, 154, 156,
regulations, 14, 30, 64, 67, 72, 94, 103, 117–118, 130
reserve (nature), 19–20, 140, 168
responsibility, 34, 42–43,73, 108, 113, 121, 123, 129, 141, 147, 149, 160, 168
rights (animal), 15, 21, 40, 117, 119

sanctuary, 49, 68, 78, 104, 106, 117, 126

safari(-s), 4, 46, 57–59, 61–68, 78, 133, 140, 169
satisfaction, 30–31, 71, 88
science/ scientific, 3, 5, 13–14, 19, 75, 77, 81, 91, 99–100, 102, 117, 128–129, 132–133, 140, 142–144, 146–147, 149–150, 170, 172
science fiction, 3, 13, 102, 128, 132
selfie, 4, 57–59, 61–68, 169
service/ services, 17, 19, 35, 45, 63, 71, 91, 114–117, 127, 131–133, 136–137
Snowpiercer, 13, 18
social media, 58, 63–68, 71, 91, 93–95, 160, 169
sport, 126, 140
Star Trek, 102
sustainability, 2, 3, 9, 16, 20, 29–36, 87–90, 113, 116, 121–123, 156, 158, 161–162, 168, 172
sustainable development, 14

technology/ technological, 2, 4–5, 11–12, 14, 20–21, 36, 47, 49, 52, 62–63, 78, 91–92, 95, 103, 114, 116–123, 127, 133–137, 140, 147, 150, 155, 156–159, 162, 163, 168, 171–172
Thoreau, 151
Tilden, 85, 170

utopia/ utopian, 4–5, 58, 65, 95, 113, 115–116, 122–123, 126, 169–170, 172
Urry, 10, 11–12, 18, 20, 167

value, 10, 31–34, 41–42, 44, 48–50, 64, 76, 95, 99–101, 105–107, 115, 117, 127, 131–132, 136, 154, 157–158, 160–163, 170, 172
virtual reality (VR), 2, 15, 20, 47–49, 80, 91–93, 120, 162

welfare (animal), 4, 15, 29–30, 40–42, 46, 48–49, 52, 54, 61–62, 64–65, 69, 75–76, 78, 82–83, 113, 114, 123–25, 144, 173
WWF, 1, 40, 146, 149

zoo, 4, 21, 41–42, 44–45, 47, 49–51, 71, 73–74, 76–81, 105, 126, 140–144, 147, 150, 169–171

For Product Safety Concerns and Information please contact our EU Authorised Representative:

Easy Access System Europe

Mustamäe tee 50

10621 Tallinn

Estonia

gpsr.requests@easproject.com

www.ingramcontent.com/pod-product-compliance
Ingram Content Group UK Ltd.
Pitfield, Milton Keynes, MK11 3LW, UK
UKHW021943200326
4879IPUK00004B/69